Oliver Roland Ingersoll

Smallest Ship that ever Crossed the Atlantic Ocean

Log of the Ship-rigged Ingersoll Metallic Life-boat

Oliver Roland Ingersoll

Smallest Ship that ever Crossed the Atlantic Ocean
Log of the Ship-rigged Ingersoll Metallic Life-boat

ISBN/EAN: 9783337320478

Printed in Europe, USA, Canada, Australia, Japan

Cover: Foto ©berggeist007 / pixelio.de

More available books at **www.hansebooks.com**

Price,]   [15 Cents.

Smallest Ship

That Ever Crossed the Atlantic Ocean:

.

LOG

OF THE

SHIP-RIGGED INGERSOLL METALLIC LIFE-BOAT,

"RED, WHITE AND BLUE,"

ACROSS THE ATLANTIC OCEAN AND ENGLISH CHANNEL.

Tempestuous Weather—
They Meet a Sail which Declines to Speak—A Whale Introduces
Himself—A Vessel that Did Speak—On Beam Ends—A Shark with
an Open Countenance—Land Sighted—Death of Fanny—Exhibition in Crystal Palace, London—Arrival in Paris—An
Interview with the Emperor Napoleon—Boats,
their Varieties and Uses, &c., &c.

NEW-YORK:
BUNCE & COMPANY, PRINTERS,
No. 350 Pearl Street.
1870.

INTRODUCTION.

WHILST science has her ardent votaries, genius her brilliant scholars, commerce her generous advocates, and agriculture her enthusiastic admirers, the progress of the age in which we live, illumined, brightened, and exalted by the investigation of the scientific mind, fostered and encouraged by the enterprize of an active, intelligent people, spreads a halo of undiminished beauty around the path of the philanthropist, and invests his pursuits with an interest which kindles within our hearts the germ of benevolence, and turns the current of our thoughts in a channel of noble and generous ambition.

Who does not feel his bosom swell with enthusiastic pride, as he contemplates the path that science has marked out? The laws of gravitation, the mariner's compass, the Hoe press, the electric telegraph, although conceived and formed by great minds, at periods remote from each other, yet forming the links of a mighty chain which was to regulate the motion of the earth, steer the frail bark of commerce, and girdle the world in an instant with the thoughts of man. Science, with renewed sagacity, points out new paths of wealth for adventurous commerce. And yet our country has scarcely unfolded half its hidden treasures; and while we enjoy the blessings of peace, the mountains and our verdant valleys will unfold new stores of wealth for the efforts of genius and untiring industry. The iron arms of progressive civilization stretch across our continent, encircle our inland seas, penetrate our granite hills, and divide the prairies of our far western lands. The massive forms of our steamships part the waters of our distant but friendly neighbors; thousands of sails, burdened with peace offerings, make their way to other shores—while the stars and stripes of a proud, powerful, and respected nation float, in "red, white and blue," upon the breeze of the most distant climes.

The almost boundless extent of our sea-girt coast, the extensive shores of our lakes, and the far stretching courses of our mighty rivers, present a field for the exercise of these noble feelings, in devising and executing plans for the preservation of human life exposed to the perils of shipwreck and disaster. There too often fatal occurrences call aloud for the strong arm of the law to interfere for the correction of abuses; whilst the melancholy testimony of the enormous number of noble lives lost appeals in earnest and urgent tones to the humane owners and agents of vessels to furnish themselves with the means which have been provided for the security of the lives of the passengers committed to their charge. Alas! how many sad hearts now mourn the loss of father, mother, sister, brother, protector or dear friend—buried in the ocean deep—that might even now have been walking the earth, but for the miserly, dastardly conduct of some ship owner or ship company. Nor are government officers, whose duty it is to protect the gallant men of our navy and army, exempt from a terrible responsibility. These men go forth with their lives in their hands for our protection. Surely, all precaution that human ingenuity could suggest should be exerted to protect them from being engulphed in the ocean's wave in case of storm, collision, or fire. With a sad heart, memory still cherishes the awful fate of those who perished in the ill fated Oneida and the City of Boston.

This little volume gives the most minute account of the voyage of the Iron Life-boat " Red, White and Blue," known as an Ingersoll Metallic

Life-boat. In addition to the letter written by the Chief of the Cabinet of His Imperial Majesty the Emperor Napoleon, appointing an interview, will be found a letter from our Minister to France, General Dix; also, a few of the opinions of the press in France, Great Britain and the United States, but only to a limited extent. Enough was published in Europe alone, both in prose and poetry, and in almost all languages, to have filled a large volume. These disjointed pieces, though, serve to give a *disinterested* and full account of the voyage and its results, if carefully and attentively read. This boat was first introduced to the public in the year 1860. It rapidly acquired favor with steamship owners and travelers. In 1865, a 26 feet boat was exhibited at the Fair of the American Institute for competition, and received the honor of the first prize, a Gold Medal. The identical boat then exhibited was afterwards decked and rigged full ship rig, christened the "Red, White and Blue," winged her way across the tempestuous Atlantic Ocean, at exactly the same time that another Iron Ship—the Great Eastern—was forging her way across the same ocean in an opposite direction, laying the telegraphic cable, which, among its first dispatches, announced the safe arrival of the Red, White and Blue at Hastings—one the largest ship, the other the smallest in the world. Neither John Bull nor Brother Jonathan, we hope, had cause to blush for the behavior of their representatives. The unthinking part of the community looked upon the "attempted voyage of the Red, White and Blue" "as a foolhardy enterprize."

But never was a voyage across the Atlantic made more in the interest of science and humanity than was that very voyage. To the inventor, it was one of the dreams of his life to construct a Life-boat that could live in all weather. To him a Life-boat was one that could live when the stoutest ship might die. Does the shipwrecked mariner with his passengers leave the ship when it can float? Do they trust their lives to the *Life-boat*, excepting when the ship is going down or in flames? Then, should not, a Life-boat be capable of crossing old ocean in its gloomiest moods, and, "like a thing of life," ride the waves, defy the hissings and lashings, and, spite of their horrible uproar, dance joyfully along with its precious freight?

It was designed to have the boat *compete* at the "Exposition Universelle" at Paris; but this, unfortunately, was lost sight of by Captain Hudson in his interview with the Emperor Napoleon, and the boat was entered for *exhibition* only, and was thus, we think, deprived of the privilege of bringing back a second Gold Medal. This was a great disappointment.

The little ship is now in New York. It is proposed soon to visit our southern coast round Cape Horn, and surprise the citizens of the golden shore of San Francisco, reserving for a future time a sail around the great inland seas of Ontario, Erie and Michigan, and down the mighty Mississippi to New Orleans. It may be among the possibilities that *a trip around the world* will finally be determined on. The Red, White and Blue can accomplish it with the same ease that she crossed the wild Atlantic, and the terrible English Channel.

To ROBERT HENRY HUNT, proprietor and editor of "Hunt's Yachting Magazine," London, whose recent sudden death has been announced, I am particularly indebted for the kind and whole-souled manner in which he received Captains Hudson and Fitch upon their arrival in England, and for the great interest he manifested in every thing pertaining to the Red, White and Blue, whose exploit, he said, "would live in the history of maritime adventure as long as the ocean rolls." Peace to his ashes! A nobler and more liberal gentleman was not known to the nautical world.

NEW YORK, April 7, 1870.   OLIVER ROLAND INGERSOLL.

THE LOG.

(*Verbatim.*)

Monday, July 9, 1866.—Mr. Fitch (and friends,) at 8·30 A. M. took ship out of Whitehall Slip, and sailed her down to Red Hook Point, preparatory to the steamer coming down to keep us company towards the Light Ship, with a party of friends to see her off. During the morning, took in and stowed stores, bread, &c. At 11·30, I went to custom house to clear; took out a register, the clearance in ballast, and the bill of health; got crew list and articles, and went through all the forms of a ship of 1,000 tons. At 1 P. M. got through at the custom house, and proceeded to Whitehall, where the steamer Silas O. Pearce was in waiting since 11 A. M., and all very impatient. 1·30, cast off and started after the ship; winds light from N. N. W., with light rain—overhauled ship Red, White, and Blue, opposite Quarantine, Staten Island; took them all on board the steamer, staid about half an hour taking leave of friends, and took a drink; myself and Fitch then went on board the ship—all the rest left her, and towed us down towards the Light Ship. Coming down the bay, our dog Fanny (which was given us by our friend, Mr. Rickhow, who also gave our preserved meats) fell overboard. A boat from the steamer picked her up and put her on board; the poor animal keeps crying, but is getting reconciled to her new home. Steamer kept on down the bay, we having fore and main topsails and jibs set, the weather looking to be clearing, and winds very light from S. W. At 4 P. M., got down to the Light Ship and cast off; hauled in the tow line—steamer laid by us and gave each of us three cheers, with a tiger, and also the ship. During this ceremony there was many a white handkerchief waving in the breeze, and from thence to the bright eyes of our warm hearted friends, who all prayed for our safety. She then proceeded up towards the city. Five set mizen topsail. 7·30, wind N. W. and fresh. 9 P. M. lost the lamp; a barque passed, steering eastward; also a ship standing for the Highlands. Midnight—Highlands bearing W. b. N. 1-2 N. 20 miles, from which I take my departure. Thus ends this day, civil time; sent down mizen topgallant and royal yards.

Tuesday, 10*th.*—Begins with fresh winds and heavy swell from N. E. At 3 A. M. Fitch called me out, took in jib, fore top staysail, and main and mizen topsail; going very well by the wind, with only fore topsail; pump ship; could not get much water out of cabin floor. 4 to 8 A. M.—4·30, spoke pilot boat Alexander T. Stewart; wanted to know how she worked; told him very well. Several sail in sight bound various ways. 6, set main topsail; 8, managed to get some breakfast, and fed the dog. 8 to 12, meridian—this morning, stowing away provisions afresh, and making more room. 10, pilot boat William H. Aspinwall spoke us. 11, set fore staysail for fore topmast-staysail; and clearing the cabin.

No observation. Cloudy and hazy.

(5)

12 to 4 P. M., nothing of account transpiring; heavy swell and winds fresh. 6 P. M., got ready and set fore trysail; swell going down, and ship steadier. 6 to 8 P. M., wind and sea moderating, but looking black and threatening around the horizon; lightning to westward; tried to pump out, but could not, ship rolling about too much. No cooking has been done yet, no regular meals; used one can of chicken. 8 to midnight, moderating, light winds and cloudy, nothing transpiring; set signal light.

Wednesday, 11*th*.—Begins from midnight to 4 A. M., with light winds, and broken clouds, and clearing away with blue sky, stars coming out, swell going down, 3, set mizen topsail. 3·30, set fore and mainsail; hove log, 1 1-2 knots. 4 to 8 A. M., several sail in sight, and moderate. 7, set cross jack. 7·30, lighted the kerosene stove, and made coffee—the first warm drink or any thing hot yet. 8 to 12, meridian, detached clouds, blue sky, a ship following after; about 10, she kept off, found she could not fetch us. Sun got past the meridian before we could take her. No observation.
Light northeasterly winds from noon to 4 P. M. From 4 to 6 P. M., light airs, and heavy swell from the east. 6 to 8, calm and no steerage. 7. spoke the schooner Pequonic of Bridgeport, hence from Boston towards Philadelphia. 8 to midnight, calm. 9·30 P. M., light airs from the westward; hauled up the mizen and main courses, squared the yards; set signal light.

Thursday, 12*th*.—From midnight until 4 A. M., light westerly winds and fine. 4 until 6, light winds and clear blue sky; sailed through a tide rip; set main topgallant sail and royal. 8 to noon, light westerly winds, and weather fine. At 9·30, set the fore topgallant sail and royal. All this watch Captain Hudson employed in drying his clothes, books, bedding, &c., and clearing out the cabin. Saw a bark steering S. E.
12 to 4 P. M., moderate breezes, with broken clouds and blue sky; saw some small quantities of gulf weed; the air rather damp. Fitch employed in getting the bed and other articles below. 4 to 6 A. M., getting things ready, and Captain Hudson cooked supper, and done the washing up of dishes; had some of the best mutton soup we ever eat of the kind out of cans. 6 to 8, fresh breezes and clear sky; small sea heaving; the ship doing well, and carrying the royals; going 7 large by the log; set signal light; lost sight of the bark. 8 to midnight, strong breezes and sea making up; carrying the following sails: fore topmast staysail, foresail, topsail, gallantsail and royal; on the main—main and topsail, top gallantsail, royal; mizen topsail. 8, set signal light.
This day expended one can mutton soup; and one can of beef for the dog, as she must not be forgot; she takes it out in sleeping.

Friday, 13*th*.—From 12 to 4 A. M., strong breezes and clear sky, with a heavy sea heaving; but the ship rides it well, shipping small quantities of water. By the looks of the water we are in the gulf stream. 4 to 8, fresh gales and clear, heavy sea running. At 8, called the watch and furled fore and main top gallant sails and royals, shipping some water, but doing very well. 8 to meridian, fresh gales from westward, clear blue sky, heavy sea heaving, ship taking in some water and running very well, better than I expected, considering, when she is in the hollow of the sea, the sails are almost becalmed. Not able to cook any thing this morning. For these twelve hours one mile is allowed per hour east for heave of the sea.
12 to 4 P. M., fresh gales, with high topping sea. When in the hollow, the topsails almost becalmed, saw a school of flying fish; shipped large quantities of water, running into the cockpit from thence below. Captain

Hudson bailed out fourteen buckets of water from the cabin; wet bed and other articles. Wet me completely at the helm. 4 to 6 P. M., found our watch stopped at 5; got wet with Fitch, and not able to set her going, having rusted, we now have to go by sun rise, sun set, and meridian, and morning and afternoon sights, for determining our time of day. Fresh breezes, heavy sea running. 6 to 8, fresh breezes and clear sky, heavy swell, but going down; bent main trysail. 8 to midnight, winds moderating some, sea going down, clear sky. Not able to cook any thing this day. This last 12 hours current E. b. N. in Gulf Stream, one mile, half mile for heave of sea. 8 P. M., set signal light.

Saturday, 14th.—Midnight to 4 A. M., moderating and clear, with heavy swell, but going down. About 2 A. M., a small barracouta jumped on board. Shipping some water, nothing more transpiring; sunrise, 3·46. 4 to 8 A. M., light breezes, swell heaving, looking black and cloudy in the N. W.; light thunder in the distance; hauled up mainsail and squared yards. 8 to meridian, light winds, cloudy, looking threatening, with distant thunder, but got nothing from it; latter part, light winds and baffling, ship not having steerage way on her, put her round several times. Opened a box of can turkey, and stowed them, and making a fresh stowage in the hold, got out the bedding and other articles of clothing to dry, pumped out two buckets of water from the hold, done various other small jobs. These twelve hours one mile per hour E. b. N., for current of stream, half mile heave of the sea. Opened a can of mutton soup for breakfast. No time for cooking, having to take observations at different times to find the time of day.

12 to 4 P. M., light airs with passing clouds, swell moderating, but looking heavy to N. E.; employed in fitting tiller ropes, overhauling clothes, stowing away bedding, and other small jobs. 4 to 6 P. M., fresh breezes and clouds, sudden flaws; furled foresail and mainsail and mizen topsail; ship going under fore and main topsails, jib and fore topmast staysail, saw a full rigged brig steering east pass about four miles to northward; heavy swell making up. 6 to 8 P. M., fresh breezes with clear sky, heavy sea, making up from N. E.; it has changed very soon. 7, a bark passed about five miles to northward, steering east. No cooking this afternoon, too rough for the kerosene stove; shipping some water. 8 to midnight, moderating, but heavy swell, clear sky, pleasant weather; a small flying fish two inches long flew on board this watch. Gulf Stream current twelve miles for these twelve hours. That cock pit of ours is a very bad place, cramped up just high enough to take the hips and make us both sore—cramps our knees. It is the hardest place on board; the rest is bad enough.

Sunday, 15th.—Midnight to 4 A. M., winds moderating, swell going down, fine and clear weather. 3·30, set fore and mainsails, and mizen topsail, nothing else transpiring. Sunrise, 3·46 A. M. 4 to 8 A. M., fine, pleasant light airs, clear sky. 8, cooked coffee and can of chicken; saw a bark astern to westward, steering easterly. 8 to meridian, fine and pleasant, with very light airs from E. to N.; swell heavy. Saw a sail to southward, nothing transpiring. The loss of our watch, by rusting with salt water, is a very serious drawback to us, as we have to go by sunrise, sunset, meridian, and fore and afternoon sights to determine our time. Midnight is the worst, as that is mostly guess work.

Current E. b. N., twelve miles from Gulf Stream.

Meridian to 4 P. M., calms, light airs, and baffling all round the compass, with clear sky. About 4 P. M., concluded to try and speak our companion the bark; light airs; got her headed towards him to get his longitude, we having no chronometer. 4 to 6 P. M., light breezes easterly, making little way towards the bark, about two miles distant. Set our ensign, which he answered, but hanging down was unable to make out his nationality. He wore round to the north, and kept off from us, *evidently not wishing to speak us*, I suppose, for fear we wanted something. I cannot say too much against that captain's humanity, whoever he may be, that would pass a small ship, with only two men in, and 500 miles away from land, without desiring to speak her, if even he could do nothing. *I leave him to his conscience.* The One that directs all will give him his reward. 4 to 8 P. M., light breeze, and clear and smooth sea; set the spanker. 8 to midnight, moderate breeze and steady; set signal light; the bark coming up astern, and hauled his wind, passed half a mile a-weather of us to gain advantage of ground, so to prevent me from speaking him, if I wanted. He need have no fears on that ground. Finished one keg of water, ten gallons, and opened another.

Monday, 16th.—Midnight to 4 A. M., clear and pleasant weather, blue sky, wind freshening, smooth sea. At 2 A. M., saw a sail ahead and going east, leaving us. Sunrise, 3·46 A. M. 4 to 8 A. M., fine and pleasant breezes, with passing clouds; nothing of any import transpiring; swell heaving up; not able to light our kerosene stove, so nothing to be cooked. 8 to meridian, fresh winds and pleasant passing clouds, and sea making up. Took altitude of sun; found time 8·45 A. M. This morning doing several small jobs. Not many lazy times on board this ship. Shipping some water.

Ther. at noon: Air, 91°—Water, 81°.

Found ourselves to south by observation of where we ought to be. Current must have set south-easterly, as it is known not to be always steady in the stream. Going under three topsails, fore and mainsails, jib, fore topmast staysail and spanker.

Meridian to 4 P. M., strong breeze, passing clouds, and smoky in sun; took in spanker; heavy sea making up, and shipping some water. Fitch drew off three gallons of water from the cask in the hold for use, took bed out to air, and put below again not quite dry. Another small flying fish jumped on board. 4 to 6 P. M., breeze freshing up, and hauling west; squared yards, and hauled mainsail up; heavy clouds in west bearing a threatening appearance. No cooking done to-day; opened a can of mutton soup. 6 to 8 P. M., breezes moderate and cloudy; set signal light. 8 to midnight, fine breezes, with dark clouds and overcast, and heavy sea heaving. Another flying fish jumped on board—killed itself instantly by striking the deck. Under three topsails, fore and mainsails, jib and fore topmast staysail. Hauled up farther to northward to-day to get nearer the gravel bank of Newfoundland. Tried our compass by the north star; found her correct; no attraction from the iron, being an iron boat, and iron all around and underneath the compass, which is rather strange.

Tuesday, 17th.—Midnight to 4 A. M., moderate and fresh breezes at intervals, with heavy dark clouds. 2 A. M., braced in and set mainsail, heavy sea heaving. Another flying fish jumped on board. Sunrise, 3·46 A. M. 4 to 8 A. M., moderate, and clearing off. Saw a large school of flying fish, and large quantities of gulf weed. 8, took altitude of sun to determine our time of day. Still no time on board. 8 to meridian, moderate, fine and pleasant, with broken clouds; sea going down. Saw several schools of flying fish, and some gulf weed. 10, set cross jack and spanker. Allowed for this last twenty-four hours, twelve miles current easterly for the stream; nothing of any moment transpiring. Going under three topsails, three courses, spanker jib, and fore topmast staysail.

Ther.: Air, 92°—Water, 81°.

Meridian to 4 P. M., light winds, pleasant weather and blue sky; swell heaving. Employed in fitting leading blocks to topsail halyards, top gallant halyards, and fitted a bumpkin forward for fore tack to lead to bows; too narrow. At 2 P. M., hauled up cross jack and squared the yards. 4 to 6 P. M., moderate breezes and clear sky; has the appearance of freshing. Took altitude of sun 4·30, to determine time of day; drew out two gallons water from the hold; set flying jib. 6 to 8 P. M., cloudy, and moderate heavy swell. Sunset 7·44 P. M. for time. 8 to midnight, fine breezes and cloudy; furled spanker. Nothing transpiring, only both very sore from sitting down in the cock pit; it is a very hard place. 8, set signal light.

Wednesday, 18th.—Midnight to 4 A. M., fine breezes and cloudy; heavy sea heaving. Another flying fish flew on board. Nothing more transpiring. 4 to 8 A. M., moderate breezes, and heavy clouds passing; heavy sea heaving. 6, set mainsail. Sunrise, 3·42 A. M. 8 to meridian, light breezes, fine and pleasant; hazy weather. 9, hauled up mainsail, squared yards, took sight for time at 8·15 A. M. Took a dry breakfast this morning; no cooking, ship rolling too much for the fire in the kerosene stove. Allowed for gulf current and heave of the sea east, true, twenty miles for the last twenty-four hours. Opened one can of turkey for breakfast. The dog Fanny sick, will not eat.

Ther.: Air, 88°—Water, 77°.

Meridian to 4 P. M., light winds and pleasant weather; clear sky; swell heaving, and ship rolling some. 2 P. M., set mainsail and spanker; nothing transpiring. 3·45, took altitude for time of day. 4 to 6 P. M., fine and pleasant. 4·30, set main top gallant sail and royal, squared yards and hauled up mainsail; furled spanker; swell heaving and wet some clothes drying. 6 to 8 P. M., moderate; took supper. Nothing cooked to-day; ship rolling so much cannot keep any thing on the stove. Sunset 7·18 to guide our time. 8 to midnight, moderate breezes and swell heavy; clear sky. At 11·30, *struck a snag on our port bow*, or some other substance, which completely stopped her headway, or it *might have been one of the rocks* marked on the chart as being doubtful position; but it is doubtful if any such do exist in the gulf stream. Called the captain; but that was unnecessary, for as soon as the shock was felt he was on deck, but could see nothing. He immediately ascertained if ship was leaking, and found she was not. On the instant our lonely position came over us, and our feelings can better be imagined than described; enough to make stouter men than us to anticipate the worst—for on very dark nights ships running large, striking any thing in the water, might instantly go down before they could do any thing to save themselves.

Thursday, 19th.—Midnight to 4 A. M., strong breezes and dark clouds heaving from westward, and overcast; heavy sea making up; shipping some water. 2 A. M., a flying fish hit me on my left cheek, but did no damage, and fell in the cock pit; and the dog Fanny amused herself with it for a little while. 4 to 8 A. M., moderate breezes and passing clouds. Sunrise, 3·42 A. M. Nothing more transpiring. 8 to meridian, moderate breezes, clear sky, heavy sea heaving; took out the bed to dry. Some water got through the companion way. Allowed for gulf current E. S. E. true, twenty miles. No fire lighted this morning. Opened one can of chicken.

Ther.: Air, 89°—Water, 79°.

Meridian to 4 P. M., fresh breezes and dark heavy clouds and overcast, with heavy sea heaving from westward; bailed large quantities of water out of the cabin; deck leaks under the gunwale, wet the wearing clothes and found them mouldy; had to wash and dry them; some of them are completely spoiled. Also took the bed out, but found it impossible to dry it, being cork not dry through; concluded to throw it overboard. Got out two gallons of water from the hold. 4 to 6 P. M., furled main top gallant sail and royal. The weather has a threatening appearance, dark and cloudy. Took in clothing, and not dry. 5·26 A. M., took an altitude for time. 6 to 8 P. M., furled jib and flying jib; strong winds and heavy sea heaving, and cloudy, and shipping water, some going in cabin. Sunset 7·18 for time. 8 to midnight, moderate, with strong winds in flaws, and presenting very threatening appearance. Took in mizen topsail at 8 P. M., and set signal light. No fire has been lighted to-day, ship not steady enough.

Friday, 20th.—Midnight to 4 A. M., moderate winds and squally, and very dark clouds and threatening. 2 A. M., hauled up foresail; heavy showers of rain. 3, set foresail and mizen topsail; continued rain during remainder of the watch. 4 to 8 A. M., the same weather continued, with constant rain. Sun obscure. Got no time this morning; have to go by guess work. 8 to meridian, light winds with squalls; dark and cloudy, with passing showers; looking very threatening. 9, took in mizen topsail. About 10, wind shifted sudden to N. E. About 11·30 A. M. furled fore and mainsails, set fore and main storm trysails; wind freshing up, and sea running cross ways, and making up from N. E. This day, cabin and every thing in it wet again; continually damp; impossible to keep any thing dry. Opened one can of turkey. Have to slack up some of the standing rigging to keep mast from breaking. Current allowed E. by S. true, eighteen miles for stream. No observation this noon. Our whisky gave all out to-day.
    Ther.: Air, 72°—Water, 70°.
    Meridian to 4 P. M., strong winds north easterly; dark heavy clouds and threatening heavy sea making from that quarter. 1 P. M., took in main topsail. 2, furled fore topsail, going under fore topmast stay and fore main trysails; making very little headway. Fitch bailed out eight buckets of water from cabin; and a box of bread in tiers, the lower tier of about ten pounds, was spoiled. Threw it overboard. Spilled most of a jar of butter. Every thing damaged; cannot keep even this journal dry. Opened another box of crackers. 4 to 6 P. M., fresh breezes, and cloudy, and heavy sea; lying under storm trysails; doing nothing but drifting. 6 to 8 P. M., weather the same, nothing new. 8 to midnight, squalls and threatening weather, heavy clouds. Put out signal lamp; heavy sea heaving; shifting some water; moon is clouded; latter part moderating. No observation throughout this day. No cooking has been done.

Saturday, 21st.—Midninght to 4 A. M., breezes moderate and sea going down; clear sky overhead; stars out. 3 A. M., put out wet clothing and set the topsails, fore, main and mizen, jib and spanker. At 3·30, wind hauling; tacked ship to N. E.; furled fore and main trysails. Sunrise, 3·46 A. M., coming out pleasant. 4 to 8 A. M., moderate breezes, sea going down, coming out clear with some clouds; set fore and main sails; shipping a little water. 8 to meridian, fine and pleasant, with sea heaving; shipping some water; nothing transpiring. Noon, under three topsails, jib, fore topmast staysail, fore and main sails, and spanker; fresh breezes. Opened one can mutton soup. Current allowed E. S. E., twenty miles. No fire lighted this morning.
    Ther.: Air, 70°—Water, 69°.
    Meridian to 4 P. M., fresh winds and dark heavy clouds, with light

showers and heavy sea; shipping some water, not much; every thing wet and damp. 4 to 6 P. M., strong winds and broken clouds; has the appearance of a cold winter's day. Bailed out four buckets of water from cabin, as it runs from the hold; pump out one gallon of water from second keg, which finishes that. No cooking done this day, ship pitching and rolling so. 6 to 8 P. M., fresh winds, dark clouds, and threatening heavy sea. Sun set 7·14 P. M., but not visible for time. 8, furled cross jack and spanker. 8 to midnight, fresh winds, very cloudy, heavy sea; put out a red signal light, our only one left now. Nothing more transpiring.

Sunday, 22nd.—Midnight to 4 A. M., fresh gales, with dark heavy clouds and passing showers of rain; heavy sea heaving. At 4 P. M., furled jib and hauled up mainsail. 4 to 8 A. M., strong gales, with showers and heavy clouds; took in mizen topsail; set fore storm trysail; sun obscure. Sunrise, 3·46 A. M., shipping some water. 8 to meridian, strong winds continued, and weather the same; took in fore and main topsail, and set main storm trysail. Ship under fore and main trysails and fore topmast staysail; had to slack up top gallant back stays, wet and shrunk, to keep the masts from breaking. At 10, wind dying out, set the three topsails, three courses, jib and spanker. Meridian, coming out a little pleasant and clearer; sun not obscure. Opened the third keg of water to-day. Have not allowed for any current this twenty-four hours, winds being easterly and southing in it, may not have any, as they influence it here.

Meridian to 4 P. M., moderate breezes and cloudy, but pleasant; heavy sea heaving. Fitch got his clothes out from the hold, got mildew in his valise; also dried our canvas bedding, and bailed out twelve buckets of water from the cabin and hold, 40 gallons. It is Sunday, but those things must be attended to. Busy all this watch. About noon this day, run out of the Gulf Stream. 4 to 6 P. M., moderate and pleasant, but cloudy looking, threatening to the south; getting things put back in their places; cabin floor a little dry for the first time since we are out. 6 to 8 P. M., moderate and pleasant, but cloudy; heavy sea, clearing up decks, took in spanker. Sunset for time, 7·17 M. No cooking to-day. Put out a red signal light, our only one. 8 to midnight, fresh breezes and heavy clouds; sea making up, with drizzling rain.

Monday, 23rd.—Midnight to 4 A. M., strong breezes, and dark gloomy weather, and heavy sea; shipping some water; showers of drizzling rain at intervals. Sunrise 3·43 for time, but clouded at the time. 4 to 8 A. M., strong winds and coming clearer, with detached clouds and blue sky; heavy sea. 8 to meridian, strong winds, with broken clouds and heavy sea; shipping some water, and making water in around the gunwale when it is under, or water on deck; ship under three topsails, fore and main sails, fore topmast staysail and jib. No cooking so far, this day taking dry meals.

Ther.: Air, 72°—Water, 64°.

Meridian to 4 P. M., fresh breezes and cloudy, and hazy and smoky around the horizon. About 3 P. M., saw a large green turtle, and passed large quantities of kelp weed; heavy sea heaving. 4 to 6 P. M., fresh breezes and hazy; heavy sea; shipping some water. Nothing doing to-day. No cooking to-day. Bailed out ten buckets of water. 6 to 8 P. M., fresh winds and cloudy, and hazy weather. Fitted a temporary gaff abaft on fore topmast head for hoisting the signal light to at nights, which gives it an elevation of twelve feet from water line. 7·19 M., sun set in time; set signal light. Opened 1 can of beef for the dog. 8 to midnight, strong breezes and thick fogs, with drizzling rain and very dark, although a good moon, but obscure; heavy sea, shipping considerable water. At 11·30, or about midnight, called Fitch; took in main and mizen topsails, fore and main sails and jib, and set fore trysail; blowing sharp in squalls, and wind not steady. Saw several porpoises.

Tuesday, 24*th.*—Midnight to 4 A. M., fresh winds and dense fogs, with drizzling rain and heavy sea; ship going easier, not shipping so much water. Sunrise at 3·19 A. M., but obscure. 4 to 8 A. M., fresh winds, dense fogs, and heavy sea; set main topsail. 6 A. M., set jib, foresail, mainsail, and mizen topsail. Not able to see through the fog over a hundred yards; sun is shining through it occasionally. Not able to take any altitude, but judge it eight o'clock by the height of the sun. 8 to meridian, moderate and fresh breezes, with heavy sea and dense fogs; hauled up mainsail; fog has shrunk the standing rigging very tight; had to slack topgallant rigging, fore and aft, to keep spars from being broken. The late rains and fogs are mildewing all the sails; they being only of cotton drilling, and set not furled, begin to look bad. Sun obscure. Opened one can of turkey.

Meridian to 4 P. M., moderate breeze, and very heavy sea running from S. and S. E., and dense fogs. Cannot see more than fifty yards around. About 3, passed through a strong current ripple; got in it before I could see it. The water for about sixty yards in a *fearful foam, and topping up five or six feet. Ship would scarcely steer, and was a long time getting out.* Sometimes her headway was stopped;—*resembles passing through Hurl Gate, New York, with it foaming and whirling,* and as if rocks might not be far from the surface. Also passed through several smaller ones. From that fact, and very cold, and feeling like ice to south, I conclude we were in the polar current. 4 to 6 P. M., moderate winds, dense fogs, with cold drizzling rain; heavy swells heaving from south. 6 to 8 P. M., light winds, heavy sea with thick fogs; can see no distance. Sunset 7·22 M., and obscure; weather becoming warmer; set signal light. 8 to midnight, moderate winds, heavy swell, and fog clearing some. Under the three topsails, foresail, jib, fore topmast staysail, and fore trysail.

Wednesday, 25*th.*—Midnight to 4 A. M., fresh breezes, and fog not so dense; the moon shining clear; heavy sea heaving about. A little before 4 A. M., current under the lee, wind abeam; when on to top of a sea, it brought her up, and come near going over. 4 A. M., squally, and taking large quantities of water on board; and being in the polar current, ship acting very bad, took in fore and main topsails, furled mainsail and mizen topsail, set main trysail. Sunrise, 3·38 M., obsure. 4 to 8 A. M., set fore topsail; wind becoming light, but heavy sea, and weather clearing. 8, set main topsail. 8 to meridian, fresh winds and heavy seas; shipping some water; cloudy and hazy around the horizon, with blue sky. Got meridian altitude. Passed through some slight ripples of current.

Ther.: Air, 74°—Water, 64°.

Allowed one mile per hour for the polar current setting south.

Meridian to 4 P. M., moderate breezes and heavy sea heaving, cloudy and hazy. Bailed out twelve buckets of water, about twenty-four gallons; and drying clothes, not having been dry since we left. Ship leaks considerable somewhere; set spanker. 4 to 6 P. M., moderate breezes and sea going down; doing several small jobs around. Our time is well taken up, principally taking care of provisions and clothes, to keep them from getting any worse; bailed out six buckets water. 6 to 8 P. M., fine and pleasant, and light breezes, with broken clouds, blue sky; set the mainsail, and furled main trysail. Sunset 7·22 M. for time. 8 to midnight, fine and pleasant, but hazy, with broken clouds; sea going down. 10 P. M., set the cross jack. No cooking done to-day. By appearance, getting out of current. Set signal light at 8 P. M.

Thursday, 26*th.*—Midnight to 4 A. M., moderate breezes, with passing clouds and hazy; not much sea. Nothing transpiring. 4 to 8 A. M., moderate breezes, and light fogs. and a swell heaving. Not having the sunrise to go by, Fitch thought his time up, and called me; half an hour afterwards, clearing up, I saw the sun just rising, 3h. 38m. A. M., thereby getting the best of me about one hour. 8 to meridian, moderate breezes and fogs—not very thick, latter part clear with blue sky. Lighted our' stove and made coffee, and opened one can of turkey and warmed it for breakfast. 10, set fore and maintop gallant sails.. This morning, drying our matches and lamp wick over the stove; wanted doing very bad, as not one in a dozen would light, and putting them in a jar. The first fire lighted this morning for ten days. Polar current south ten miles. Noon out of current.
 Ther.: Air, 72°—Water, 63°.
 Meridian to 4 P. M., light winds and clear sky; swell heaving. Got out clothes to dry; but not much of a drying day; air damp, although it is pleasant, especially so when the wind is fair. Also dried some of the spare sails, steering sails, &c. Bailed out five buckets water. 4 to 6 P. M., fine breezes, pleasant and clear; furled fore trysail, and took clothes down; took in spanker. 6 to 8 P. M., fresh winds with clouds, and a heavy swell heaving. Sunset 7·21 for time. A dense fog setting in, took in cross jack and put out signal light. 8 to midnight, fresh breezes and thick fog; heavy swell heaving from S. W. This day, carrying fore and main top gallant sails. Nothing further transpiring.

Friday, 27*th.*—Midnight to 4 A. M., strong breezes, and squally; dark heavy clouds, and fog clearing; heavy sea making; shipping much water, and lee rail under most of the time. 3 A. M., called Fitch out, and took in fore and main topgallant sails, and hauled up mainsail; looking very squally. 4 to 8 A. M., fresh breezes and cloudy, heavy seas. Sunrise, 3·39. About 7, more moderate; set the mainsail. 8 to meridian, strong breezes and clear, with broken clouds and heavy sea heaving; shipping some water, and doing some jumping. Made some coffee this morning. Noon pleasant. Opened one can of chicken. Under three topsails, fore and mainsails, jib, and fore topmast staysail.
 Ther.: Air, 78°—Water, 72°.
 Meridian to 4 P. M., moderate, and gentle breezes, with blue sky and detached clouds. Got out our canvas bedding and some clothes to dry, as they have been wet since we have been out, and not dry once. Saw a flying fish, rather unusual in these latitudes. 4 to 6 P. M., moderate and pleasant, hazy around the horizon, heavy sea swelling; bailed out eight buckets water, and took down bedding and clothes; ship goes along nicely, with fair wind, throwing sprays quite often. 6 to 8 P. M., mild and pleasant breezes, clear sky, sea going down, hazy around the horizon. Sunset for time, 7·21; set our signal light forward. 8 to midnight, light . breezes, and pretty smooth, clear weather. 10 P. M., set main topgallant sail and royal. Ends pleasant, with a bright moon.

Saturday, 28*th.*—Midnight to 4 A. M., fresh breezes, with squalls; some broken clouds, blue sky. 1 A. M., clewed down main royal. 4 to 8 A. M., moderate breezes, blue sky and hazy, but pleasant and smooth sea. 6 A. M., set main royal. Lighted the kerosene stove, and made coffee for breakfast at 8. 8 to meridian, breezes freshing up and sea making, clear sky. This morning, captain broke water breakers out of the hold, and took two in cabin in place of two empties and stowed them; broached another, which makes the fourth water keg; now found some of them not full by two gallons; filled them out of that one; filled water can for drinking, to be able to fill it soon with salt water; filled three empty ones with salt water to keep ballast good, as she begins to feel getting lighter below. Sea making up, and shipping some water.

Ther.: Air, 72°—Water, 69°.

Meridian to 4 P. M., strong breezes, with clear sky, and heavy sea making up, and shipping some water. Fitch employed in finishing filling the water kegs, and putting every thing below again. 2, clewed down main royal. The ship feels the little extra weight below of the three kegs of water—more steady. 4 to 6 P. M., wind increasing; furled main topgallant sail and royal; furled the mizen topsail and cross jack; ends fresh gale and heavy sea, clear sky. 6 to 8 P. M., strong gales, with heavy flaws at intervals; blue sky, with broken clouds; seas making heavy; had to keep her before them, when large ones came along, to keep them from getting on board. Sunset 7.21 for time; set signal light. 8 to midnight, strong gales and heavy seas continued; had to keep her before them when heavy ones came, but making her course good when past. Running under fore and main topsails, fore and mainsails, jib and fore topmast staysail. No cooking in latter part.

Sunday, 29th.—Midnight to 4 A. M., strong gales and dark clouds to N. W.; clear sky, and very heavy sea running—having to keep her off when large ones come; shipping some considerable water. At sunrise, 3.39, wind shifted to west. Called Fitch out to square yards and take in the mainsail and jib. Wind moderating, but sea making from west, and making a bad cross sea. 4 to 8 A. M., fresh gales, and heavy cross sea; running under fore and main topsails, foresail and fore topmast staysail; sea too dangerous to set more sail; wind appearing to moderate, with clouds around the horizon. Not able to light fire this morning. 8 to meridian, winds fresh and moderating, clear sky, detached clouds around the horizon, sea going down, the same sale continued on the ship, shipping some water. Have allowed for this twenty-four hours twenty miles for heave of the sea; 1-4 point is allowed southerly for griping to windward.

Ther.: Air, 79°—Water, 69°.

Meridian to 4 P. M., winds moderating, and heavy swell heaving; clear sky, hazy around the horizon; set mainsail and mizen topsail; bailed four buckets of water out of cabin (12 gallons); set spanker. 4 to 6 P. M., light winds with blue sky, with broken clouds; heavy swell heaving. Drew off two gallons water out of the keg in the hold. 6 to 8 P. M., light breezes and heavy swell, clear sky. Sunset 7.24 for time; set signal light. No cooking to-day. 8 to midnight, calm and clear sky, with light clouds around horizon to westward; hauled courses up and spanker.

Monday, 30th.—Midnight to 4 A. M., calm. 2 A. M., light breezes from S. W.; set courses and spanker. At 3 A. M., a large whale came alongside, or within thirty feet; kept ship away from him. Called captain at sunrise, 3.36 for time. 4 to 8 A. M., light breezes with clouds; swell gone down. Lighted stove and made coffee this morning. 8 to meridian, fine and pleasant, with moderate breezes. 9 A. M., set fore and main topgallant sails and royals. Hudson washing his mildewed clothes and drying them; scarcely any thing dry as yet. Current has set southerly twenty miles this last twenty-four hours. Opened one can of turkey, and one can of beef for the dog.

Ther.: Air, 73°—Water, 63°.

Meridian to 4 P. M., moderate breezes, and cloudy. Fitch washing and airing his mildewed clothes, and clearing the anchor rope; so full of kinks cannot do any thing with it; had to tow it overboard to take out the turns. 4 to 6 P. M., fresh breezes and cloudy, not looking very good weather; furled the fore and main top gallant sails and royals. Always have to send them on deck to do this, and send them up again. 6 to 8 - P. M., breezes freshing up and cloudy, swell heaving. Sunset 7·23 for time; set the signal light. Finished the fourth keg of water. 8 to midnight, moderate gales, and sky overcast, and dirty looking weather. Nothing transpiring.

Tuesday, 31st.—Midnight to 4 A. M., fresh breezes, dark, gloomy, and overcast, with some small drizzling rain, not much. At about 3·50, took in spanker, hauled weather clew of mainsail up, and squared the yards. 4 to 8 A. M., fresh breezes and cloudy, with fogs and drizzling rain. At 8, made coffee; had some trouble with the stove not burning good. Opened fifth keg of water. 8 to meridian, moderate breezes, cloudy and overcast, with some drizzling rain. About 9 A. M., set main top gallant sail and royal; the fore will not draw very well if set. Ship is under foresail and topsail, mainsail and topsail, and top gallant sail and royal, mizen topsail, and spanker, jib, and fore topmast staysail. Latter part sun coming through the clouds; got observation; swell heaving. Opened one can turkey.
Ther.: Air, 68°—Water, 62°.
Meridian to 4 P. M., strong breezes and cloudy; saw quantities of sea weed and a piece of plank. Nothing has transpired. 4 to 6 P. M., fresh breezes, cloudy, and overcast; sea making up, shipping some water. 6 to 8 P. M., strong breezes, cloudy and hazy weather; furled spanker— latter part furled main top gallant sail and royal; wind increasing; set signal light; sun obscure. 8 to midnight, fresh breezes, overcast and gloomy weather; latter part moderate. The water sparkles very much; the water almost light when sea breaks towards midnight. Several seas in succession came rolling along, carrying the ship with them with great velocity, so that the sails were hard aback, and the yards square. *I have never seen the like before with any thing at sea.*

Wednesday, August 1st.—Midnight to 4 A. M., fresh breezes and fogs; small rain at intervals, with heavy sea heaving. 4 to 8 A. M., winds moderating, and cloudy, with thick fogs. Sun is obscure at rising for our time. 7, wind hauling, squared yards, and hauled up the weather clew (starboard) of mainsail. 8 to meridian, moderate breezes, heavy swell; shipping some water; fog is continuing—very damp, almost like rain; sun coming out at noon and clear. Got meridian altitude. Opened one can of mutton soup; finished second box of crackers—opened another box of crackers. Twelve miles allowed east for heave of sea, one-half point southerly is allowed on the course for griping and sea heaving her up.
Ther.: Air, 72°—Water, 64°.
Meridian to 4 P. M., moderate breezes, cloudy and gloomy weather, heavy swell heaving; ship under the three topsails, fore and main sails, jib and fore topmast staysail. Nothing transpiring. 4 to 6 P. M., moderate and cloudy, heavy sea, square yards. 5, parted mainsheet on port. Opened one can of turkey. 6 to 8 P. M., moderate, cloudy, gloomy and fogs. Sunset obscure, 7·22; set signal lamp. 8 to midnight, moderate, and heavy swell, with thick fogs; moon don't show through it. No cooking done to-day.

Thursday, 2nd.—Midnight to 4 A. M., moderate winds, thick fogs, with fine drizzling rain; heavy swell continuing. 3, hauled up mainsail. Sunrise 3·38, obscure. 4 to 8 A. M., light winds, with dense fogs, and swell continuing. Sun obscure. Drew two gallons of water from the keg in hold. 8 to meridian, light winds with fogs; swell continuing. 9, set fore and main top gallant sail and royals. 10, squared yards and hauled up port clew of mainsail; set spanker. 11, took in spanker; latter part sun coming out, but not very clear. Got meridian altitude. Finished one box of herrings. Most every thing spoiling. Current twenty-four miles south. By observation, ship has gone considerable south; differs twenty-seven miles from yesterday's reckoning.

Ther.: Air, 71°—Water, 62°.

Meridian to 4 P. M., light and moderate breezes, fogs and cloudy weather, and a heavy sea heaving. At 4, looking breezy; furled fore top-gallant sail and royal. 4 to 6 P. M., fresh breezes, and swell continuing, with thick fogs. Quite a number of ripples on the water. so that it has the appearance of a current. Opened one can of chicken. No cooking this day. 6 to 8 P. M., fresh breezes and fogs, and heavy swell continuing, with drizzling rain, cold and disagreeable. Sunset at 7·22, but obscure; set the signal light. 8 to midnight, breeze moderating, sea going down, some fog still continuing, with fine drizzling rain; nothing transpiring. Only the cock pit, where we steer, being compelled to sit, it is with difficulty we can keep awake sometimes.

Friday, 3rd.—Midnight to 4 A. M., fresh breezes and cloudy, and fogs, with swell heaving; going along nicely. Sunrise 3·38, but obscure. Nothing transpiring. 4 to 8 A. M., wind moderate, with thick fog; swell heaving not so heavy. Drew off one gallon of water from keg, and opened one can of chicken. 8 to meridian, moderate winds and fogs, with swell heaving. 9, set fore top gallant sail and royal. 10, saw a school of porpoises; did not stay long around. This morning, Hudson setting things to rights in the cabin, and overhauling his valise for mouldy clothes; was successful in finding some shirts that way inclined, and books also. Got meridian altitude, but not very good one, on account of fog not rising much. Have not experienced any current by the reckoning this day.

Ther.: Air, 68°—Water, 62°.

Meridian to 4 P. M., moderate breezes, and hazy and cloudy weather; fogs cleared off. Drew off two gallons of water from keg, and other small jobs; swell continuing, but going nicely; very near all sail set. Bailed out two buckets of water, swashing around. 4 to 6 P. M., moderate and cloudy weather, and overcast; heavy swell heaving. 6 to 8 P. M., moderate breezes and overcast, hazy, with fine drizzling rain. Sunset at 7·22, but obscure; set signal light; clewed down fore and main royal. 8 to midnight, overcast, with moderate breezes and swell heaving, and drizzling rain, with fogs in latter part; nothing transpiring. No cooking done this day.

Saturday, 4th.—Midnight to 4 A. M., moderate and light winds, swell heavy, fogs and fine rain latter part; squared yards, wind hauling to northward. 3·30, set the royals. Sunrise at 3·38, but obscure for our time—so we have at those times to go by guess work; a serious inconvenience to be without the time at sea. 4 to 8 A. M., light winds, cloudy and hazy; thick weather, with small rain; wind dying out about 5 A. M. and coming N. E.; braced up on port tack, and set fore and main sails and spanker; weather continuing. 8 to meridian, moderate breezes, with cloudy and gloomy weather, and hazy, with drizzling rain. 10, clewed down the royals, heavy swell heaving, winds baffling. At meridian, tacked ship to north, going under easy sail; hauled courses up; sun out very dim; altitude not to be depended on.

Meridian to 4 P. M., moderate breezes and dark gloomy weather, and looking threatening, with drizzling rain and a heavy sea, and heaving to westward. 1 P. M., furled the courses; 2, furled top gallant sails; 3, set fore and main storm trysails. 4 to 6 P. M., light winds and heavy swell; weather has same appearance. 6 to 8 P. M., light winds, not steady; heavy sea heaving; ship rolling some; a bright horizon; clearing away in N. W. 7, saw a bark in N. E. standing east; first sail seen in twenty days. Sunset 7·17, and first time seen set for ten days. No cooking this day. 8 to midnight, calm and light airs, blue sky with clouds and stars out. Nothing transpiring.

Sunday, 5th.—Midnight to 4 A. M., calm, clear, and chilly; sea going down. Nothing transpiring. Sunrise at 3·43 for time. 4 to 8 A. M., calms and light airs, and very fine and pleasant. Got out my mouldy clothes to dry, and had to dip some of them in salt water; the best airing day we have had so far. In the morning, opened one can of chicken. Made coffee this morning. 6 A. M., set all sail except flying jib and mizen top gallant sail and royal. 8 to meridian, light winds and pleasant weather, with blue sky and broken clouds. 9, sighted a sail to S. E.; kept for her. At 10·30, she came down to us—the bark Danish Princess, of Yarmouth, N. S.—asked if we wanted any thing. She hove to; we stood round under her stern. Gave us a bottle of rum; very good for wet days, as we had none; also an old white light and two newspapers, the *Irish Times* and *Freeman's Journal*. The Danish Princess was eleven days from Dublin, bound for Quebec; broke the pole end of our jib-boom under her counter; reported longitude 22° west, but not to be depended on, as they had no observation for several days, so I will keep my own.

Ther.: Air, 68°—Water, 62°.

Meridian to 4 P. M., fresh breezes with broken clouds; took in mizen topsail and spanker; sea making up; shipping water and getting damp; took in clothes, put them away; I threw some overboard in a bag; my English and French colors were in also; did not know it until too late. 4 to 6 P. M., fresh breezes continuing and cloudy, with sudden flaws of wind; furled fore and main top gallant sails and royals; heavy sea making up; shipping considerable water; furled spanker. 6 to 8, same weather continuing; heavy sea, shipping water. Sunset 7·17, but obscure. 8 to midnight, strong breezes and heavy sea, dark cloudy weather, shipping water and pitching some.

Monday, 6th.—Midnight to 4 A. M., strong gales and heavy seas, shipping much water; filled the cock pit full once, latter part; furled mainsail, squally. 4 to 8, same weather continuing, very heavy seas, shipping water. 8 to meridian, winds moderating, but seas heavy; ship will not keep out of their way; shipped two seas over the stern, filled the deck—cock pit escaped this time; ship rides the sea well; those seas she only took the fragments, as her sharp stern split them, but she took quite enough. Saw a sail to eastward, steering north-easterly; breeze up again. Opened one can of chicken.

Ther.: Air, 64°—Water, 64°.

Meridian to 4 P. M., light breeze and cloudy, threatening (with squalls) a very heavy cross topping sea, and shipping a great deal of water; sea very dangerous. At 4, bailed out twelve buckets of water which had run below from the cock pit; furled mainsail and mizen topsail. 4 to 6 P. M., winds moderate, but not able to carry any sail, as it is squally; very heavy sea, and ship laboring considerable and shipping water; clewed down main topsail. About 5, Fitch was forward on starboard bow; a blind sea took ship on port quarter, and hove her on her *starboard beam ends—let* go fore topsail halyards, *she righted in about half a minute.* We have carried sail pretty hard, but never saw her do that before; the sea was the cause, as only fore topsail and foresail, jib and topmast stay sail were on. 6 to 8 P. M., fresh breezes and the same sea, very dangerous to us. Passed a ship steering S. W., with only her light sails in. Sunset 7·17, but obscure. 8 to midnight, fresh winds and dark gloomy weather, not able to carry any more sail to keep her out of the way of the sea, as it is squally. During this day, have had to steer various courses to keep out of the way of the sea from boarding her and serving us the same again, and also winds not steady, veering around. No cooking done this day.

Tuesday, 7th.—Midnight to 4 A. M., more moderate, with clouds passing, sea more regular. At 3 A. M., called the watch (Hudson); set main topsail and starboard clew of mainsail; sea going down. 4 to 8 A. M., winds moderate, with passing clouds, not so much sea. 5, set mizen topsail, going nicely; neither of us have slept this night past; our cabin is miserable, constantly wet; our bed and clothes all wet; cannot change them, or no use to do so. This last twenty-four hours we would have been very bad off, had it not been for the bottle of rum we got from the Danish Princess, on Sunday last; kept our life in, as we cannot do any cooking. 7, took in mizen topsail, squally. 8 to meridian, moderate, heavy clouds passing, and considerable sea going yet. At 8·30, shipped a very heavy sea, which hove the ship flat on her starboard beam ends, the yard arms were in the water, and all that side of the foresail and topsails; only fore and main top sails, foresail and jib and fore topmast staysail were on at the time; let go and clewed down the top sails—no small job just then, but that was our only chance—she came back again in about one minute, although the time looked an hour. We are not able to carry much sail, on account of the weather being puffy and squally, to keep her out of the way of the sea. Passed a bark steering S. W., at 10 A. M. Eighteen miles S. E. is allowed for heave of the sea.
Ther.: Air, 69°—Water, 62°.
Meridian to 4 P. M., moderate, set mizen topsail, coming on cloudy and squally with rain about; furled mizen topsail, clewed down fore and main topsails; squalls passed over rather heavy. Set fore and main topsails again; wind moderate but veering, not so much sea. Fitch bailed out six buckets of water from cabin that ran in during the morning, and filled empty keg with salt water. Began on the sixth water keg. 4 to 6 P. M., light breezes, with passing clouds and heavy sea. Set mizen topsail. 6 to 8 P. M., moderate, cloudy, sea more regular. Sunset 7·12, but obscure. Set signal light, and took in mizen topsail; the weather don't look like carrying much sail with the sea heaving. 8 to midnight, moderate, with dark clouds and squalls, and bearing a threatening appearance. Heavy sea still running. Many times during this day, had to keep ship before the large seas, as they heave in a south-easterly direction.

Wednesday, 8th.—Midnight to 4 A. M., strong breezes arise, very heavy squalls, with dark clouds and rain. From 1 A. M. to about 3, a strong blow of continued squalls, with rain; clewed down fore and main topsails, let her scud under the foresail and fore topmast staysail; did not call the watch, as every thing leads aft. Sunrise 3·46, but obscure. Set fore topsail again. 4 to 8 A. M., fresh winds and very cloudy, and threatening appearance. About 7.30, a heavy squall struck us; clewed down fore topsail; called the watch, and hauled tight the gear; the fore and main sails are furled by that means. Squall lasted about half an hour. 8 to meridian, strong breezes, with dark cloudy weather. About 9, set fore topsail. Very heavy seas running from westward. Got meridian altitude, but rather dim. Opened can of chicken. One mile per hour allowed to S. E. for heave of the sea.

Ther.: Air, 60°—Water, 60°.

Meridian to 4 P. M., heavy gales, with dark cloudy hazy weather, and passing showers of rain, with very heavy sea running, but regular. 1 P. M.. furled fore topsail, running under foresail and fore topmast staysail. Ship scarcely able to keep out of the way of the sea, laboring heavy, and shipping considerable water; sent down fore and main topgallant and royal yards; heavy squalls attending throughout. 4 to 6 P. M., strong gales, with squalls, thick and heavy showers of rain. Sea running cross from S. W. and N. W., making bad work of it. 6 to 8 P. M.. gales moderating, but dark and cloudy, sea not so heavy; set fore topsail. Sun set 7·14, but obscure; set signal light. 8 to midnight, moderate gale, sea going down, but running in almost every direction, and striking the ship from all quarters, pitching and rolling very bad. During this day, had to keep the ship before the sea at times when it looked the worst. No cooking done to-day.

Thursday, 9th.—Midnight to 4 A. M., moderate and cloudy, hazy, sea going down. 3, set main and mizen topsails. Sunrise 3·46 for time. 4 to 8 A. M., light winds, and dark heavy clouds flying, and apparently clearing away from westward. 6, set jib, mainsail and spanker; 8, hauled up spanker; light drizzling rain. 8 to meridian, light winds, hazy and cloudy weather; mist all driven off. 10, hauled up mainsail and squared yards. Doing some general ship work—not many lazy times. Got meridian altitude, but dim. Current twenty-four miles allowed for heave of the sea, S. E. by E., true. Opened one can of mutton soup.

Ther. : Air, 67°—Water, 59°.

Meridian to 4 P. M., light breezes, with black clouds rising in N. and W., with a heavy swell heaving S. E. Got out our canvas bedding to dry, and put below again. Bailed five buckets of water out of cabin; cleaned cabin. A school of porpoises are playing around the ship. 4 to 6 P. M., gentle breezes, with passing clouds. 6 to 8, moderate breezes, hazy and cloudy, swell heaving. Sunset 7·14 for time. Set signal light. 8 to midnight, strong gales, with heavy squalls and passing showers. At midnight, furled main and mizen topsail, jib and mainsail; heavy gales blowing, and very rough seas; settled the fore topsail half way down, and find that enough. No cooking to-day.

Friday, 10th.—Midnight to 4 A. M., strong gale in the first part, with heavy black clouds passing, latter part moderate; have to keep ship before the sea at times, and never mind the course; running under half of fore topsail, foresail and fore topmast staysail. Sunrise 3·46 for time. 4 to 8 A. M., gale moderating, with a heavy sea heaving to S. S. E. 8 to meridian, fresh breezes, and detached broken clouds; heavy sea; ship laboring heavily, and shipping considerable water. We are wet the most of the time. Sun out bright at meridian. Twenty-four miles allowed for heave of sea to S. S. E., true. Opened one can mutton soup.

Ther.: Air, 68°—Water, 62°.

Meridian, fresh gales, passing clouds. 3, moderating; set fore, main and mizen topsails; heavy sea, shipping some water. 4 to 6 P. M., moderate, broken clouds, chilly weather, sea heavy. 6 to 8 P. M., same weather. Sunset 7·14 for time. Set signal light. 8 to midnight, moderate, heavy sea heaving to south throughout. Nothing transpiring. Had to keep the ship off course, before the sea, many times during the day. No cooking done.

Saturday, 11*th*.—Midnight to 4 A. M., gentle breezes with passing clouds, seas continuing, but going down; set mainsail. 4 to 8 A. M., light breezes and passing clouds; swell heaving. 8 to meridian, light breezes, passing clouds, swell heaving; sent up fore and main top gallant and roya' yards; set main top gallant and royal; set up main topmast back stays, and other small work. Meridian altitude puts us thirty-three miles south of Ushant, and our longitude nearly up. Finished sixth keg of water, and began on seventh; finished the third box of crackers; opened another, most of them mouldy; opened two tins of turkey. Twelve miles allowed south for heave of sea.

Ther. : Air, 70°—Water, 60°.

Meridian to 4 P. M., moderate breezes and coming fresh, hazy and overcast. At noon, found my reckoning up for longitude; kept north to get in latitude of Ushant, and ran east to sight it. 3·30, fresh breeze and sea making up; furled jib, main topgallant sail, royal, mizen topsail, and mainsail; drizzling rain. Filled empty keg in hold with salt water. 4 to 6 P. M., blowing a fresh gale, drizzling rain and overcast; furled main topsail. 6 to 8 P. M., strong gale, with very heavy and dangerous sea; our latitude being up, kept off more to east to suit the sea; shipped some very heavy ones, filled deck and cock pit; furled fore topsail, running under foresail and fore topmast staysail; weather same; set signal light. 8 to midnight, strong gale, overcast, dark and gloomy, with dangerous seas, shipping water. Have to keep her before the seas.

Sunday, 12*th*.—Midnight to 4 A. M., moderating, with squalls, cloudy and thick weather, heavy seas continuing; shipping water, bailed some out of cabin. Saw a large shark alongside. 4 to 8 A. M., strong breeze and sea the same, and shipping water. About 8, while captain was getting his breakfast, shipped a very heavy sea between main and mizen rigging, *which completely knocked her on her beam ends*, and me (Fitch), *up to my neck in the water*; I let go the helm then, seeing it was of no use, and grasped the mizen mast, to keep myself from going overboard. After the sea passed over her, doing all the damage, taking some small things with it, and filling the cock pit (holds about two barrels), and half filling the cabin, *she righted again*, making the *fourth* time she has been on *her beam ends since we left*. 8 to noon, moderate, sea not so heavy; got meridian altitude, not very clear; seems inclined to break away. We have every thing wet now, bed and clothes; cabin is very miserable, wet and damp all the passage; bailed out about forty buckets, but no chance to dry any thing.

Ther.: Air, 60°—Water, 54°.

Meridian to 4 P. M., light winds, cloudy, and heavy sea; set jib, the three topsails, mainsail and spanker; saw a bark steering S. W. 4 P. M., sounding sixty fathoms, white sand. 4 to 6 P. M., light winds and heavy swell; saw a ship steering S. W.; a steamer south; bailed out six buckets of water. 6 to 8 P. M., moderate, cloudy, and a heavy swell. Sunset 7·15, but obscure; set signal light. 8 to midnight, moderate, with strong breezes at intervals. 10, furled mizen topsail and spanker, clewed down fore and main topsails; set them again; same throughout. No cooking this day.

N. B.—From our position at noon, found we had overrun the reckoning, as the longitude puts ship close up to French shore, and the soundings at 4 P. M., sixty fathoms, white sand, and latitude 48° 56′ N., puts ship in position of Island of Ushant (France), bearing south twenty-seven miles, from which I take its longitude 5° 5′ W. as my departure, having overrun my reckoning sixty miles, from New York to Island of Ushant; not very bad in a distance of 3,300 miles.

Monday. 13*th.*—Midnight to 4 A. M., moderate and steady breezes, with clouds and overcast; saw quantities of kelp. 3, set mizen topsail, cross jack and spanker. Sunrise obscure, 3·45 A. M. 4 to 8 A. M., fresh breezes, not steady. 6 A. M., hauled up cross jack, saw large quantities of kelp; swell going down. 8 to meridian, moderate and pleasant, with broken clouds, not promising weather. 10, saw a steamship steering east. Noon, sounded; got forty-three fathoms, white sand and shells; have not sighted land yet. Opened two cans of turkey.

Ther.: Air, 72°—Water, 58°.

Meridian to 4 P. M., light winds, cloudy and threatening, sea moderate, several sail in sight beating down channel. 2 P. M., furled cross jack, standing in to make the land about Start Point. 4 to 6 P. M., light winds and drizzling rain; saw a bark to leeward. 6 to 8 P. M., light winds and cloudy, standing down to head the bark off; she hove to, set American colors; Nettie Merryman, of New York, Captain H. A. Rawlins; came to under his lee quarter; he put on board two bottles of brandy and a broken white signal lamp; from Havre, with passengers bound for New York; gave our position, Start Point, bearing north thirty-five miles, and N. E. to the Bill of Portland; kept off for that place at 7·15 by his time. 8 to midnight, fine breezes, overcast, with showers of rain, in latter part; furled mizen topsail.

Tuesday, 14*th.*—Midnight to 4 A. M., moderate and cloudy. At 2 A. M., wind shifted to N. W., gybed ship on port tack. 4 to 8 A. M., fresh breezes, clouds and blue sky. 7 A. M., set mizen topsail, mainsail and spanker. Saw a large fleet working down channel. 8 to meridian, fresh breezes and passing clouds. 9 A. M., spoke Heron brig, H. V. Troop, of Liverpool, N. S., bound for Messina; reported the Bill of Portland N. E. 10 *A. M., made the Bill of Portland, N. N. E.* Passed through a very large fleet of shipping of all classes and nations working down channel. Meridian, Bill of Portland bearing N. N. W., twenty miles distant.

Ther.: Air, 70°—Water, 58°.

Meridian to 4 P. M., moderate breezes, passing clouds. 2 P. M., exchanged flags with a Russian bark standing in for Poole Point, bearing N. by E. 4 to 6 P. M., fresh breezes, several sail in sight, strong ebb tide S. W. Poole Point N. by W. 6 to 8 P. M., moderate breezes. 7 P. M., Needles Light bore north. 8 to midnight, moderate and clear. 10 P. M., abreast of St. Catherine's Light, Isle of Wight, bearing N. by W., going along with fine steady breeze. A brig passed, steering east.

Wednesday, 15*th.*—Midnight to 4 A. M., moderate and fine, several sail beating down channel. 2 A. M., hauled in more for the land. 4 A. M., abreast of Newhaven. 4 to 8 A. M., moderate and fine breezes; saw a bark-rigged steamship steering S. W. 6 A. M., Beachy Head bore E. N. E.; set cross jack. 8 to meridian, fine breeze, with clouds and hazy, keeping along the land. 11 A. M., abreast of Beachy Head, one mile distant; set fore and main royals, top gallant sails, and flying jib. Meridian, Eastbourne abeam N. W., about two miles distant; a number of vessels lying at anchor off there. Opened two cans of turkey, and one of mutton soup.

Ther.: Air, 62°—Water, 60°.

Meridian to 4 P. M., moderate breezes, clouds, and blue sky. 3 P. M., passed a revenue cutter, which saluted us, but did not board; several sail going various ways; abreast of Hastings, and close to, spoke several boats. 4 to 6 P. M., fresh breezes and cloudy; quite a number of boats came off from Hastings; got some papers from them, and learned for the first time that the Great Eastern had successfully laid the Atlantic Telegraph Cable.

S to 6 P. M., the same weather, and boats around all the time, saluting the little Yankee ship most vigorously. 8 to midnight, moderate and calms. 9 P. M., made Dungeness Light E. N. E.; calms. Not making much way, several fishermen around. Steering courses along the land, and following it around the bends to keep out of the tide.

Thursday, 16*th.*—Midnight to 4 A. M., fresh breezes and moderate, several sail passed, steering various courses up and down channel. 2 A. M., Dungeness bore N. by W., one mile distant. 3 A. M., abreast of Dover. 4 A. M., South Foreland bore N. E. two miles. 4 to 8 A. M., moderate and light winds. 7, abreast of Deal, several boats came off; wanted to know if I (Hudson) did not want to take a man for pilot. I want to take her all the way, so that no one can say they took her any part. 8 to 12 A. M., moderate and fine, with clouds; sailing along the land. About noon, Broadstairs bore N. E.; kept up for that place; a steamer came off and spoke us. Passed Ramsgate; a boat came off there also to report us. Meridian to 4 P. M., fresh breeze, and coming on to blow very heavy from W. S. W., so could not carry any sail; rounded South Foreland, and wind being so, had to beat up to Margate. Coming on very heavy could not carry any sail, ship being on her beam ends, and several boats came off from Margate to see her; found we could do nothing, flood tide also; the boat Jessie of Margate, Captain Thomas Watler, came off and towed us in; furled sails, made several tacks to fetch in. About 3 P. M., got inside the pier in the basin, and the crowd on the pier gave many lusty cheers for the little ship that had so successfully braved the Atlantic. The rest of the day many people came off, and *we went on shore* and put up at the Hoy Hotel, Mr. Stevens. During the remainder of the day being entertained by the several people around, and many visitors also.

Friday, 17*th.*—This day, strong winds from N. W., and nothing favorable towards making a start. Have to lay still. Many people coming to see the little ship from all quarters, all around the country; some say she never came across—but if they had the privilege of reading this log, I think they would alter their opinion. Winds the same during the latter part of the day.

Saturday, 18*th.*—This day fine and pleasant, winds westerly. About noon, light winds N. E. 2 P. M., flood tide, floated out, made all sail and stood on the course for Sheppy Island; very light winds easterly; a number of small boats accompanied the ship out for two miles. The dog Fanny is very sick; she has been well taken care of and had a warm bath, but is very feeble; had several fits during the night. 4 to 6, moderate and fine, but hazy, making very little way. 6 to 8 P. M., fine weather and clear, winds very light from S. W.; doing very little, ebb tide setting; furled fore and mizen top gallant sails and royals; could carry them easy, but too much sail on for working quick. 8 to midnight, passed Herne Bay about nine o'clock, and still bearing about the same at midnight, S. W.; ebb tide setting; passed the Girdler Light Ship, flood tide making up at midnight; quite a number of vessels going up. The dog has had a number of fits this watch.

Sunday, 19*th.*—Midnight to 4 A. M., moderate, very light winds, still under way but making very little way; daylight, got up abreast of Sheerness; several vessels around; nothing transpiring. The dog still getting worse; can do nothing for her. 4 to 8 A. M., light winds easterly; not doing much; several steamships pass up and down. 6 A. M., up with Southend; the steamer Londonderry, Captain White, ran alongside, and very kindly asked if I wanted a tow, which was very thankfully accepted ; gave him a line and furled sails. About 5·30, by the time of a sloop close by, the dog Fanny could hold out no longer, and *died at my feet* in most fearful agony, and had fits very frequent; I could do nothing that I knew to save her ; after being with us all the time, and now after all danger is past, to lose her is almost too much to think of. At about 8, one of the New York and London line of steamers passed and saluted; asked him to report me on his arrival forty-one days out to-day, and been capsized four times on the passage. 8 to meridian, moderate breezes easterly; going up towards Gravesend about 11 A. M., the steamship anchored to wait for tide ; ship was besieged with boats coming alongside and tearing things in general; still lying astern of her, received a visit from Captain White of the Londonderry, and returned on board with him, and accepted of his hospitable invitation to dinner. 1 P. M., she got under way ; we cast off, and took on board Mr. Charles Thomas Marshall, boatman of Gravesend, to pilot the ship up to Greenhithe; set topsails, fore and mainsails, and jibs; sailed up; went on shore. On board the Coast Guard Ship got permission to lay her alongside, where she will lay to paint and clean her a little, and until a proper place can be procured up in the city for her.

Ship making her passage in thirty-four days from port to channel, thirty-eight from New York to Margate—the first port put into from stress of weather, and New York to Gravesend, when boarded by customs officers, and received a clear pass, forty days and sixteen hours.

The log is kept in civil time, same as on shore, so no change in respect to nautical time is necessary. In reading these remarks, if not stated, that in nearly all instances when any work, making sail, furling sail, and other things, that one had to do it, the other steer.

The third stove has not been very much use to us; we have not been able to light more than twelve days during the whole time, either from water flying over her, or being so unsteady, and have not had warm coffee more than twice; out of five pounds of coffee only used one pound, the rest spoiled with the water; our bread in the latter part of the passage, from dampness, is all spoiled, having to throw nearly half of each box away.

6 P. M., hauled ship astern of the Coast Guard Ship at Greenhithe, and made fast; anchor out ahead, so that she can be cleaned before proceeding to the city.

Summary of Log.

Date	Course.	Miles Dis.	Diff. Lat.	Dep.	Lat. D. R.	Lat. by Obs.	Variation.	Diff. Lon.	Lon. in.	Winds.
July										
10	S.E.¼E.	42	25	31	39·58		¼W.	41	73·17	W.N.W. to N.E.
11	S.E.¾E.	49	30	39	39·22		¼	51	72·29	E.N.E.
12	S.75°E.	43	118	41	39·10	39·35	½	53	71·36	E.N.E. to S.W.
13	E.¼S.	168	8	168	39·27	39·24	¾	216	68·00	W.S.W.
14	E.¼N.	139	7	138	39·31	39·32	1	178	65·02	W. to N.E.
15	E.S.E.	63	23	58	39·09	39·01	1	75	63·47	N.E. to S.W.
16	E.¼N.	92	5	91	39·00	39·06	1¼	118	61·49	S.W. and W.
17	E.N.E.½E.	124	38	118	39·44	39·48	1¼	155	59·14	N.W. and W.
18	E.N.E.	104	41	96	40·29	40·31	1½	126	57·08	W.N.W. and W.
19	E.N.E.¾E.	115	28	41	40·59	41·01	1¾	148	54·40	W. to S.W.
20	E.N.E.½E.	99	29	94	41·30		2	125	52·35	W. to N.E.
21	S.E.¼S.	26	19	17	41·12	41·15	2	23	52·12	E. to S.E.
22	N.N.E.	72	66	29	42·21	42·19	2	39	51·33	S.E. to S.S.W.
23	N.E.b.E.¾E.	119	51	107	43·10	43·13	2¼	147	49·04	S.W.
24	N.E.b.E.¼E.	109	59	93	44·12		2½	130	44·56	S.
25	E.N.E.¾E.	85	20	82	44·32	44·31	2½	114	45·02	S.
26	E.N.E.	89	33	82	45·04	45·05	2½	117	43·05	S. to S.W.
27	E.N.E.¼E.	117	37	110	45·42	45·35	2½	156	40·29	S.S.W. to S.W.
28	E.N.E.½E.	117	34	112	46·08	46·06	2½	162	37·47	S.W.
29	E.b.N.	172	36	167	46·42	46·38	2½	241	33·46	W. to N.W.
30	E.½S.	51	5	50	46·33	41·33	2½	73	32·33	S.W.
31	E.N.E.½E.	129	37	124	47·10	47·09	2½	182	29·31	W. to S.W.
Aug										
1	E.	130		130	47·10	47·17	2½	191	26·30	S.W. to N.W.
2	E.¾S.	109	15	107	46·54	46·51	2½	147	23·53	W.N.W. to W.
3	E.N.E.½E.	114	31	109	47·22	47·20	2½	160	21·13	W.
4	E.¼S.	76	5	75	47·15	47·05	2½	110	19·23	Around compass
5	N.W.	25	19	18	47·24	47·21	2½	27	19·50	E. to S. W.
6	E.½N.	109	10	108	47·31	47·28	2½	159	17·21	W. to N.W.
7	E.½S.	103	12	102	47·16	47·16	2½	150	14·53	N.W. to W.
8	E.N.E.½E.	124	33	112	47·49	47·50	2½	168	12·05	S.W. to N.W.
9	E.¼N.	109	3	108	47·53	47·53	2¼	162	9·23	W. to W.N.W.
10	E.¾S.	102	10	101	47·43	47·40	2	151	6·52	N.W. to W.
11	E.¾N.	73	15	66	47·55	47·56	2	100	5·12	W.
12	N.E.	83	58	60	48·54	48·56	2	91	3·41	W.S.W. to W.
13	N.E¾N.	69	51	46	49·45	49·40	2	69	3·56	N. to S.W.

August 14th.—At meridian, Bill of Portland bearing N. N. W., 20 miles distant, and by the position when spoke the bark Nettie Merryman on the 13th of August, Start Point 35 miles. I have overrun my log about forty miles instead of sixty, as I supposed, as I stated on the 12th to the Island of Ushant, France, allowing for the tides, of the set of them I was not acquainted with; wind W. to N. W.

15th.—Eastbourne abeam, bearing N. W., 2 miles, W. to S. W.

16th.—Towed into Margate about 3h. P. M., W. S. W.

We hereby certify that this Log is a true and authentic account of the voyage.

(Signed)

John M. Hudson, Captain,

Frank E. Fitch,

Ship—Red, White, and Blue, of New York.

From " Hunt's Yachting Magazine," April 1, 1867.

THE LITTLE ATLANTIC SHIP--RED, WHITE, AND BLUE.

THE adventurous American mariner, Captain Hudson, with his equally hardy mate, Mr. Fitch, have sailed for Paris, to add their gallant little ship to the many wonders of the Exhibition. On Sunday, the 3rd of March, an immense crowd of spectators assembled on the pier at Dover, to witness if the tiny ship was actually going to face the heavy easterly gale and mountainous sea which prevailed in the Channel. The steam tug Palmerston towed her clear of the harbor, the sea making a clean breach over her decks more than once during the service, whilst the Red, White, and Blue, was over and under like a wild duck; once cast off, however, she rose like a sea bird over the rolling waves, and, spreading her canvas wings, dashed away in gallant style for the shores of France, amidst enthusiastic cheering and waving of hats and handkerchiefs from the host of wonderers on shore.

An equally exciting scene took place when she was being towed by the Souvenir out of Caen en route for Havre. The sea was breaking furiously on the bar, and the steamer filled her deck three times; but the little ship and her daring crew went through it merry as sandboys, to the astonishment and delight of the crowd of French that lined the pier.

We understand that a similar incredibility to that expressed by many visitors to the ship, when at the Crystal Palace here, prevails amongst our French neighbors, but not to the same extent. However, Captain Hudson and Fitch can afford to smile; they have satisfied all whose opinions are worth having. We should not wonder at all if their passage across the channel ditch was as equally incomprehensible, and as obstinately denied.

We subjoin a copy of her Log to Havre, kept in civil time:

Wednesday, February 27*th.*—Launched little ship into Regent's Canal, and re-rigged same day.

Thursday, 28*th.*—Hauled out of Regent's Canal Docks; steam tug Rose took hold of our rope and towed us to Gravesend, where we let go our anchor at 1 P. M.

Friday, March 1*st.*—At 7 A. M., made fast to the British brig Delhi, in tow of the steam tug Magnetic; strong breezes from E. and N. E., with passing showers of hail throughout the day, and very heavy sea heaving in from the eastward. Little ship jumping considerably whilst in tow. About noon, off Sheppy Island. At 4 P. M., steam tug cast off Delhi—which made sail. At 5 P. M., North Foreland bearing W. N. W., about three miles, parted the tow rope in the very heavy sea still heaving; we set fore and main topsails, and foresail, and kept off west for the Downs. At 7·25 P. M., passed South Foreland. 8·30, at gun fire ran into Dover, made fast, night looking very dark, cloudy, and threatening, and the mate Mr. Fitch not well. The coast-guard did not know what to make of us— our coming in took them by surprise.

Sunday, March 3*rd.*—During the whole of this day there were strong gales from east, with very heavy seas heaving. At 10 A. M., the steam tug Palmerston towed us clear of the Admiralty Pier, when we set foresail and topsail, and kept her S. W., to get under the French land. At meridian, passed Folkestone, distant about six miles; also at same time a screw steamer hailing from Cork, bound east, and three British fishing luggers, all of which saluted us as we passed. 2 P. M., Dungeness bore N. N. W., distant about fourteen miles. Up to sunset ship doing very well, but pitching and jumping considerable, and shipping great quantities of water occasionally; the wind hauling a little to N. E. at times. At 8 P. M., sighted a revolving light on Point D'Ailly bearing S. S. E., distant about six or eight miles. We must have had a southerly current setting us in towards the coast, or should have made St. Valery or Fecamp first. At 10·30 P. M., Point D'Ailly bearing east, we made Fecamp lights about west, found ship close in shore, so hauled off west and W. N. W., to round the light. Midnight, lights abeam, strong wind, and heavy sea.

Monday, March 4th.—Commences with strong gales and heavy sea, the little vessel shipping considerable water. At 1·30 A. M., made the two lights of Cape Le Heve, bearing about S. W. by W., and sought for a third, expecting it was Fecamp. As the first light was revolving at Point D'Ailly, the next to be seen was St. Valery; but not seeing that, took Fecamp for it, as it showed the same lights, the red tide light not to be seen, although we were close in, so that about 4 A. M., when I became certain of the Cape Le Heve lights, we were too far off to fetch in under the Cape the way it was blowing, so we hauled up the foresail and concluded to let her go in towards the shore, when we made the high land between Havre and Caen. Looking for the entrance, we saw a sloop and schooner coming round the shore, also two steamers. Got through in time to follow the schooner in, and at 8 A. M., made fast to the steamer Souvenir of London, Captain French, who took us in tow to Oysterschaven, the entrance to Caen.

Thursday, March 7th.—At 8·30 A. M., the Souvenir got up steam; we made fast, and she towed us over Caen bar, and to Havre. At 11 A. M., cast off at the end of Havre Pier, and the steamer proceeded on her voyage to London. At noon, we got inside and snug in Havre Dock.

We take this opportunity of expressing publicly our sincere thanks to Captain French, of the British steamer Souvenir, for the extreme courtesy and attention we experienced from him whilst at Caen, and for his kindness in towing us from thence to Havre.

(Signed)

John M. Hudson, Captain,
Frank E. Fitch,

Ship—Red, White, and Blue, of New York.

Havre, March 10th, 1867.

[Since the above was written, we have learnt that the little vessel has arrived safely at Paris.—Ed. H. Y. M.]

FURTHER PARTICULARS.

The Red, White, and Blue was launched at 4 P. M., on the 21st June, 1866. Previous to launching, she had been on exhibition for a few days at junction of Thirteenth Street and Broadway, and was taken on this date to be launched on her briny element. While resting on the dock, her shoe or false keel was put on, three and a half inches deep, which gives her seven and a half inches keel altogether. The time of high water having arrived, she was put bows to the edge of the dock, on account of jib-boom and bowsprit being out; and the dock to the water being a drop of five feet, no ways were under her. All her spars were in, and yards aloft—no ballast of any kind on board. When all was ready, those on board left her. I (J. M. Hudson) deeming the occasion a very good one to find out if she was in any way not adapted for the projected voyage, was the only person launched in her. She was got off the trucks, and as many men as could put a hand to assist crowded round her, and one good shove shot her on to the water. She did not not take a bucketful on deck, and, considering the weight aloft, did not look to turn over, or in any way feel crank, as the people on the deck said she would, but found themselves disappointed.

For a few days the ship laid in the same place, at foot of Pike Street, East River, to get sundry small jobs done while opposite the shop. It was quite amusing to me while lying there, and at the Battery also, to hear the many thousands of people that came to see her, say "That thing will never get across!" "That captain ought to be tried for murder!" And others, "They ought to be put in a lunatic asylum for two fools!" And the poor ladies—bless them!—they were all praying for us.

As her name was not made public for private reasons, every one seemed to have a name of his own for her. The "Herald," the "Lilliputian," "Brooklyn Eagle," the "Fools' Own!" some "Captain Hudson's ship!" the "Yankee Doodle!" and many others; it was only when she made her trial trip, on the 27th June, that it in any way came out. At 11 o'clock on that day, hauled to the end of the dock and set all sail on her; cast off at 11·30, having on board six persons beside myself, and in the hold 120 gallons of water. She stood her sail remarkably well, wind being east at first, and variable afterwards, until 1 P. M., light winds set in from south, began beating down the bay as far as Staten Island; stopped there, left at five in the evening; proceeded to city, wind southerly; laid her inside the Battery to finish getting stores, &c. on board, and wait until the day of sailing, the 9th of July. Whilst lying there, many more remarks were made, similar to those above noticed; but let them have their own opinions; I had mine.

The object of this expedition is to be at the World's Fair in Paris, to show the French they have not all complete without something notorious to give the rest a contrast; so, consequently, I have to start this present year 1866, as the Fair opens in April, 1867, and cannot go that year to be in time—so I have concluded to go to England, to pass away the time until the beginning of the next year.

It must be understood in reading this Log, that in the remarks, "making and taking in sail," that there is only one on deck, and he steering, so consequently he has to call the other to do what is related, with very few exceptions, or any other work about the ship. And also that this Log is kept in civil time, the same as on shore, beginning at midnight and ending on the next midnight; but the working up the day's work goes in at noon, which makes one half of two days the latter part of one and first part of the other.

Particulars of Vessel.—Built by O. R. Ingersoll, 159 South Street, New York. Length, 26 feet; beam, 6 feet 1 inch; depth of hold, 2 feet 8 inches; from deck to keel, 3 feet.

She has a sharp stern, and has a water tight compartment in each end of four feet, and cylinders in each side go to, but do not join the compartments; she is all metallic, and completely decked over, with a small cock pit around mizen mast.

The spars made by Arthur Bartlett, 252 South Street;—dimensions as follows:—

Bowsprit, 2 ft. outside bows; jib-boom, 3 ft.; flying jib-boom, 2 ft.; pole, 10 in.; fore mast from deck is 6 ft. to top, 6 in. mast head; topmast, 7 ft. 6 in., 18 in. below top for mast head, 5 ft. hoist, 1 foot mast head; topgallant mast, 3 ft. 9 in.; royal mast, 2 ft. 6 in.; pole, 1 foot; mainmast, 7 ft. from deck to top, 6 in. mast head; the topmast, gallant, and royal, are the same proportion as the fore; mizen mast from deck to top, 5 ft. 6 in., 5 in. mast head; mizen topmast, 6 ft. 2 in., 4 ft. hoist, 10 in. mast head; topgallant mast, 2 ft. 6 in.; royal, 1 ft. 3 in., 10 in. pole; spanker boom, 8 ft. long; gaff, 5 ft.; 10 in. pole on each: fore and main lower yards, 10 ft. each; topsail yards, 7 ft. 6 in. each; topgallant yards, 5 ft. 3 in. each; royal yards, 3 ft. 6 in. each; cross jack yard, 7 ft. 3 in.; mizen topsail yard, 5 ft. 3 in.; mizen topgallant yard, 3 ft. 6 in.; royal yard, 2 ft. 9 in.; fore topmast steering sail booms, 5 ft. 6 in. each; fore and main topgallant steering sail booms, 4 ft. each.

The sails made by D. M. Cumisky, 39 South Street, are as follows:

1 jib, 1 flying jib, 1 fore topmast staysail, 1 foresail, 1 fore topsail, 1 fore topgallant sail, 1 fore royal, 1 mainsail, 1 main topsail, 1 main topgallant sail, 1 main royal, 1 cross jack, 1 mizen topsail, 1 mizen topgallant sail, 1 mizen royal, 1 spanker. All these are bent to set—1 fore topmast steering sail, 1 fore topgallant steering sail, 1 main topgallant steering sail, 1 storm fore staysail, 1 storm fore. trysail, 1 main trysail. She was also draughted for her spars, the setting of them, and their length, by Mr. Cumisky. The ship was fitted and rigged by Captain Hudson, and one rigger to help.

The master's department consists of one boat compass, 1 quadrant, 2 charts from Hatteras to Newfoundland, 1 chart North Atlantic, 1 chart from Feroe Isles to Gibraltar, 1 pair of parallel rulers, 1 pair of dividers, 1 weather indicator, 1 14-sec. glass, 1 log line—8 knots, 1 7 lb. lead, 40 fathoms of line. Books—library, 1 Bible, 1 Prayer Book, 1 Bowditch's Navigator, 1 Blunt's Coast Pilot, 2 Rogers' Commercial Code of Signals, 1 Nautical Almanac, 1 Masonic Journal.

Colors, &c.—1 American Ensign, 1 English do., 1 French do., 1 American pennant, 1 anchor and rope—50 fathoms, 1 water pump, 1 tin dipper, 1 piece of rubber hose. Ship has no chronometer on board. Surgeon's department—fitted by Major John T. Lane.

The preserved can meats were presented by Isaac Reckhow, 34 Summit Street, Brooklyn: 2 dozen cans roast beef, 2 dozen cans roast turkey, 2 dozen cans roast chicken, 2 dozen cans mutton soup. He also gave Captain Hudson his dog Fanny.

Stores put up by William H. Rich: 200 lbs. of bread assorted, 5 lbs. of coffee, 2 lbs. of tea, 10 lbs. of butter, 4 boxes of smoked herrings, 1 dozen cans of milk, 1 piece of smoked beef, 15 lbs.; 1 cheese, 17 lbs.; 6 bottles pickles (stolen), 1 can mustard, 1 can pepper, 1 box salt, 1 bottle Worcestershire sauce, 12 ten-gallon water kegs, 2 bottles of brandy, 1 of whiskey, 2 of bitters. Stores enough for three months.

After the life-boat's arrival in England, some few persons doubted the fact of her ever having sailed across the ocean, when the following challenge was sent to Europe, and was extensively copied by the press. It was, however, never accepted.

To the Editors of the London Herald :

Sir,—Doubts have been raised as to the Ingersoll Metallic Life-boat Red, White, and Blue having made the passage across the Atlantic. As the little boat and her brave crew were spoken in mid-ocean, after having left this port, one would think this in itself was enough; but if there are any who are not satisfied even yet, I am willing to wager the sum of 10,000 dollars in gold against 1,000, that she did cross the ocean; and, further, I will wager 10,000 dollars in gold against a like amount that Captain Hudson and Mr. Fitch can do it again. I will give the first 1,000 to the poor of London; and on the second wager, if I win, I will give one half to the poor of London and Liverpool, and one half to Captain Hudson and Mr. Fitch. This certainly is a fair offer to any and all who would seek to deprive both the men and the boat of the credit they deserve.

Very respectfully, your obedient servant,

OLIVER ROLAND INGERSOLL,
Metallic Life-boat Builder,
South Street, New York.

New York, September 24, 1866.

———

Soon after the voyage, the Danish Princess arrived in England, and her gallant captain and crew hastened to make and publish the following, which we extract from the *London Daily News,* Oct. 30, 1866, and which forever settled the question; and those who had been loudest in expressing their doubts, now became the warmest in expressing their satisfaction, coupled with regret for the pain they had caused Captain Hudson and Mr. Fitch.

At Troon, the 22d October, 1866, appeared before me, Robert C. Reid, one of her Majesty's justices of the peace for the county of Ayr, Mr. George A. Baker, master of the bark Danish Princess, of Yarmouth, Nova Scotia, from Quebec, who declares that Captain Tooker, the former master of the said bark Danish Princess, on his arrival at Quebec, reported having spoken the little ship Red, White and Blue, from New York bound for London, twenty-seven days out, under full sail with royals set. And that in overhauling the log book of the said bark, he there found recorded the day, date and position of the little ship when it was spoken. The day was Sunday; the date 5th August, 1866; and the position, latitude 47° 19′ N., longitude 22° 10′ W. He, the said George A. Baker, further declares, that in joining the said bark Danish Princess, the only individuals belonging to the former crew he found were two lads, now on board said bark, whose names are William John Norman and Phillip M'Cormick, and who make declaration as under.

(Signed)   G. A. BAKER.

Signed and declared this 22d day of October, 1866, before me at Troon.   R. C. REED, J. P.

We, the above designed William John Norman and Phillip M'Cormick, having been on board the bark Danish Princess on Sunday, the 5th day of August, 1866, on the passage from Dublin to Quebec, declare that we fell in with, and spoke, the little ship Red, White and Blue, from New

York to London, 27 days out, under full sail with fore and main royal set. The crew on board the little ship consisted of two men and a dog. They were asked if assistance was needed, to which they answered "No. All well."

W. J. NORMAN,
PHILLIP M'CORMICK.

Signed and declared at Troon, this 22d day of October, 1866, before me,

R. C. REID, J. P.

After the Red, White and Blue sailed, the following appeared in the public prints:—

From "New York Herald," July 11, 1866.

The ship Red, White and Blue, two and a half tons, Captain Hudson, from New York for London, was spoken by the pilot boat A. T. Stewart, No. 6. The wind was blowing north-east at the time, with heavy sea on. The little craft was hove to under a fore topsail, and was behaving well. She rode the waves equal to a large ship, according to the pilot's account.

Telegraph to "Brooklyn Union."

THE LITTLE SHIP RED, WHITE AND BLUE SPOKEN.

Boston, July 12, 1866.

Captain Crowell, of the steamer Norman, from Philadelphia, reports that on the 11th instant, at 6 A. M., eighty miles east of Sandy Hook, saw the little ship Red, White and Blue, from New York for London. She was going off finely under easy sail.

The following announcement was made, per Atlantic Cable Telegraph, to the Associated Press, and was the cause of many congratulations, both from private parties and the press. It was among the first telegrams that came over the cable.

GREAT BRITAIN.

ARRIVAL OUT OF THE LITTLE SHIP "RED, WHITE AND BLUE."

Hastings, August 15, 1866.

The little ship "Red, White and Blue," of two and a half tons, from New York, with two men on board (Captains Hudson and Fitch), passed here to-day. All well. Thirty-seven days' passage.

Upon the arrival of the Red, White and Blue in Paris, being the first ship that ever entered Paris from this shore, or any foreign port, the excitement was intense. All classes seemed to unite in their congratulations, from the Emperor in the Palace of the Tuileries to the humblest workman in blouse. Captain Hudson soon received the following from the Emperor:

CABINET DE L'EMPEROR— *Palais des Tuileries, le 22 Mar. 1867.*

Monsieur,—J'ai l'honneur devous informer que l'Emperor vous recevoir demain jeudi a 9 heur des matin.

Receivez, Monsieur, l'assurance de m' consideration distingue,

Le Chef De Cabinet de I. M., CONTI.

Captain Hudson, writing from Paris, states:

Last Thursday, the 23d instant, the Emperor wrote me to visit him at 9 A. M. Punctual to the minute, I was in his presence. I was privately with him one half hour. He appeared very much satisfied as well as pleased. In closing, he asked " what he could do for me ?" I told him I wanted his influence to enter the ship in the Exposition, and be allowed to make a charge. He told me to wait. He left, but in ten minutes returned and personally handed me a letter to Monsieur Le Play, Commissioner General of the Universal Exposition; and that proved to be her permit, as I received a second one from the Emperor, to go to Monsieur Le Play, as there now was a place for her. If it had not been for the Emperor, this arrangement could not have been made. "Long may he live!" Again he writes: I am now in the Exposition. The Empress has visited the ship, and was very much pleased.

General Dix, our Minister to France, also wrote the following letter:

Paris. April 22nd, 1867.

DEAR SIR,—I regret that official engagements will prevent me from accepting your invitation to be present at your reception of friends this morning; but I beg you to be assured that no one appreciates better than myself your boldness in crossing the ocean, in this age of Great Easterns, in so diminutive, yet so gallant a craft as the Red, White and Blue.

CAPTAIN HUDSON. I am, respectfully, yours, JOHN A. DIX.

Upon the arrival of the ship in London, the following letter was received:

London Crystal Palace, Sydenham, S. E., Sept. 4, 1866.

SIR,—I have to inform you that I have arrived safe with the ship Red, White and Blue, originally Ingersoll's metallic life-boat, after a very stormy passage, making thirty-eight days from New York to Margate; and I am pleased to state that she has exceeded my expectations in the weather we have had, and I pronounce her fit to live in any sea, for it is next to impossible to capsize her altogether. We have had her on her beam ends four times, masts in the water, and got her back by getting the topsails down, when the air in the side cylinders forced her up again. Nothing shifted in the hold—ballast all secured. And I do not hesitate to say, had we had a wooden boat, with the seas she has encountered, we never could have brought her across the Atlantic Ocean. And also, you are at perfect liberty to publish this letter, if you wish to do so, as I am ready at any time, and have done since I have been here, to testify to the same. I am, very repectfully, &c. J. M. HUDSON, Captain of Ship Red, White and Blue.

To O. R. INGERSOLL, Esq., South St., N. Y.

I most cheerfully concur in the above, and cannot find language to praise the many good qualities of your Life-boat. I am positive that no life-boat yet produced, but the Ingersoll, could have brought us through.

F. E. FITCH.

OPINIONS OF THE PRESS.

ENGLAND, SCOTLAND, FRANCE AND AMERICA.

From the " New York World," August 18, 1866.

LILIPUTIAN NAVIGATION.

The Trip of the " Red, White and Blue"—The Atlantic Traversed by a Life-Boat—The Smallest Ship that Floats—Her Comparison with the Great Eastern—Description of the Men and Ship—Resume of Previous Attempts, &c.

The fairy little shell of a ship, christened the Red, White and Blue, that left New York on the morning of the 9th of July, arrived in London on the 16th of August, rounding the passage in thirty-eight days. This announcement will gladden not a few of the relatives and friends of the dual crew, and relieve the curiosity, not unmixed with interest and anxiety, of the whole public, that regarded the enterprize as an American attempt, and wished, while they doubted, its success. The miniature craft was spoken on the 10th and 11th of July, well under way; but since then, no tidings had been received of her, until the cable told us she was safe on the other side, and "all well." The trip of the Lilliput vessel is an affair of not inconsiderable importance. It may fairly and contrastively complement the passages of the Great Eastern. One is so large as to have been at first esteemed unmanageable; the other is so small as to have been from the start declared not to be able to live in any rough sea. Success has dispelled the scepticism and apprehension that each gave rise to. The first is massivity made nautically available; the second is dwarfish symmetry rendered demonstratedly seaworthy. The one depends upon power and bigness, the other on staunch minuteness and agility. The one is propelled by five monstrous engines and thirty-two sails of extraordinary extent; the latter flies by the wind, has no steam, and stretches out but sixteen airy, tiny wings, that woo the breezes, and are the whole motive power.

THE LEVIATHAN AND THE MINNOW COMPARED.

An examination of the sizual and constructive differences between the ship that is larger than Noah's ark and the vessel that barely equals in extent the poetized boat of the Lady of the Lake, will exhibit some very suggestive and instructive points, that may be figured from what follows:

GREAT EASTERN.—22,500 tonnage, 680 feet length, 83 feet breadth, 60 feet depth, length of principal saloon, 400 feet: storage capacity, 19,000 lbs.; power of engines, 2,600 horse-power; diameter of cylinder, 76 inches; draft of water, 30 feet; ordinary accommodations, 4,000 persons; greatest accommodation, 10,000 persons; highest rate of speed, 15 knots; first passage, 14 days; crew, 300; total original cost, $5,000,000; height of saloons, 40 feet; width of cable, 26 inches; weight of main anchor, 2,500 lbs.

Red, White and Blue. — 2½ tonnage; 26 feet length; 6 feet breadth; 1½ feet depth; length of principal saloon, 5½ feet; storage capacity, 1,250 lbs.; power of propulsion, 2 small children; diameter of masts, 3½ inches; draft of water, 15 inches; ordinary accommodations, 2 men (or 1 woman) and a small dog; greatest accommodations, 3 men (or 1½ women) and a moderate dog; highest rate of speed, 10 knots; first passage, 38 days; crew, 2 men and 1 dog; total cost, $1,000; height of saloon, 18 inches; width of cable, ¾ inch; weight of anchor, 25 lbs.

PREVIOUS ATTEMPTS IN SMALL CRAFTS.

The little Vision will be remembered. It was simply a sixteen feet yawl-boat, converted into a ship, or rather a hermaphrodite brig, four feet ten inches in width, and two feet nine inches deep. She carried fifty yards of canvas, and her masts were nineteen feet high. Built out of wood, she was launched by her owner, builder, and attempted navigator, Mr. John Donnovan, on the 12th of June, 1864, and started three days after, ostensibly for Europe. On the 5th of July, the Vision put into Boston, in a leaky condition; but was repaired, and started on her further way. The morning of the 20th July, the incoming steamship Peruvian spoke the Vision, doing all well, in latitude 45° 10′, longitude 33° west; supplied her with provisions and water, when again she continued on her voyage. Since that date till now, nothing has been certainly heard of the little craft, and it is believed that she has inevitably gone to the bottom. Much interest was felt in her success, and for a long time the public, on both sides of the water, refused to believe her perished. Apochryphal stories have been started now and then that she had turned up, and every port, from Spitzbergen to Terre Del Fuego, has its tradition of the fate of the Vision. But it remains now settled beyond reasonable doubt, that Davy Jones has adopted the miniature vessel as his favorite pleasure yacht down below, unless the last circulated report be true, which is unlikely, and whose purport is, that the captain put into a Nova Scotia port, staid there *incognito*, broke up his boat, gave by his absence a reasonable supposition for his death, and thus, having allowed his wife time to collect a fat insurance policy made on his life, returned to New York to enjoy the proceeds. That report don't hold water. No more does the one that the Vision ran the blockade then enforced off Charleston, and smuggled in a grist of quinine to the agued rebels. The imaginative salts that delight to resurrect the Vision, might as well conclude that they will never see her again, unless they start from Heart's Content, and striding the Atlantic cable, and pulling themselves along old Ocean's bed, consult the "arrived" list of Pater Neptunus.

THE HISTORY AND THE ACTORS OF THE PRESENT ENTERPRIZE.

In the recent Fair of the American Institute in New York, a gold medal was awarded to O. R. Ingersoll, Esq., for his improved metallic life-boat, now in such general use. The boat that took that prize passed up the Thames two days ago amid the wonder and cheers of thousands of John Bulls, who, when they sent us over their big ship, never thought we would send them in return the smallest craft that ever lived in a sea. Early last spring, Mr. Ingersoll was waited on by a little, natty sort of man, five feet two in his boots, with light sandy hair, red whiskers, open features, and an eye that looked right straight ahead from its depths of deep blue. He said: "I want to rig that boat into a full three-masted ship, go over in her to Europe, and enter her for the Paris Exposition. Will you let me have her?" "Yes" sealed the bargain. The boat, already air and water tight, was furnished with three masts, sixteen feet high; a full set of sails, amounting to sixty-five yards of canvas in all; was cargoed with enough for two men for eighty days, including the rations of a poodle dog, that was to be taken for company, and to be used as a mop now and then to clean decks. The bold men, whom all thought fools, and whom success has shown as skillful as adventurous, are Captain John M. Hudson and Captain Francis Edward Fitch, both of whom "ran away to sea" in youth, and were cuffed up from cabin boys to commanders by rapid progression, the one being 42, the other 28 years of age. We described in the *The World* their departure on July 9, and can never forget the anguish with which friends and kin bade them good by, as they cheerily cut loose from us beyond the light ship. Till the log of that memorable voyage is published, it will be difficult to know the vicissitudes and the adventures through which they passed in their thirty-eight days of solitariness. Alone; thousands of miles from land, the port of destination thousands of miles yet distant, in the midst of the ocean, that might suddenly break over them in mountainous fury, in what was scarely larger than, and exactly the shape of a tray, the like of which had never ventured to sea before; stigmatized as fools by all who mourned what they believed to be their certain self-destruction; their life bound up in their boat; given barely any sleep, and compelled to constant watching; these two men have finally safely reached the other side.

From London to New York is, in round terms, 3,500 miles, and the thirty-eight days of their passage would rate their daily going at 92 3-19 miles, or some 3 9-10 miles on an average every hour. This appears slow. Remember, however, the changes and chances of weather, and that as a purely sailing ship, the Red, White and Blue was exposed to them all, and the calms and adverse winds, and their progress is much faster than most sailing ships of 1,500 tons, and even 2,500 tons burden, which seldom are not less than forty-five and often than sixty days *en route.*

THE FRUITS OF THIS SUCCESS.

One or two things may well be concludingly borne in mind about this achievement: *First.* The whole thing—the ship, equipment, the men, the idea, the result, every thing, even the poodle—is American. *Second.* The Red, White and Blue is a full-rigged ship, as much so essentially as the Warrior and the Niagara. She was conducted as a ship, not as either schooner, brig, or sloop, which would have been less difficult and dangerous, but also less creditable and wonderful in the grand working. *Third. Passengers need have small fear to commit themselves to the Ingersoll life-boats in mid-ocean when compelled to leave the ship. There has been made in this life-boat no changes from all others of the same build,* except such alterations as were entirely external. This vessel has safely weathered very rough seas, because the Great Eastern, that was shuffling off the immortal coil at the same time, circumstantially reports an extremely severe passage. While this increased the danger and difficulty of the voyagers, and the apprehension of those of us that vividly remembered their situation, it is a tribute to the men and the vessel, which, now that their safety and reputation are assured, will be referred to with pride, where it was but recently spoken of with anxiety.

From the "Liverpool Express."

The ship-rigged boat Red, White and Blue, of two and a half tons, which has made such an extraordinary passage across the Atlantic from New York, and put into Margate a few days since, has arrived in the river off Greenhithe.

From the "New York Journal of Commerce," September 5, 1866.

*   *   *   *   *   *   Had the trip been made over a smooth sea, it would have been sufficiently marvelous; but the log of the boat, just published, shows that throughout the entire voyage the weather was unusually boisterous, with very heavy seas. Hardly had the Bank been passed, when Neptune, apparently angered at the contempt with which her crew treated his domain, sent a succession of heavy gales, which four times threw the vessel on her beam ends. And when these forces proved ineffectual to retard her progress seriously, fogs were commissioned, when off Portland, to confuse and confound. But in spite of all these impediments, the sea-king was beaten, compelled to witness her safe arrival at the port, where she was welcomed with the plaudits of thousands of delighted spectators. If the old proverb, "looking for a needle in a haystack," has been deprived of its force by the recent discovery of a slender wire in the mid-ocean, the story of the "Three Wise men of Gotham, who went to sea in a bowl," may indeed have been a veritable fact.

From the "London Era," September 22nd, 1866.

THE "WEE CRAFT," RED, WHITE AND BLUE.

"Sma' we craft" so brave and true,
Tiny bark, Red, White and Blue,
Welcome to our sea-girt isle,
Where honest Freedom loves to smile
 On manly enterprize; while in her breast,
Perhaps, the smaller is the greater guest.
Thou little dot upon wide ocean space,
How didst thou dare to show thy saucy face
Where others, of gigantic build and form,
Quail 'neath the breeze and tremble in the storm?
And *sink*, alas! whilst, seeming reckless, thou,
In the sea trough, aloft, and now alow,
Still bounding onward, like the white seamew,
Sleptst on the wind, calmly, as others do,
Pillow'd on down, such as 'neath kingly heads
The velvet hand of gentle Science spreads!
But whisper me, as in old classic tales,
Breath'd there no spell to waft thee through rough gales?
Heard'st thou no syren, floating by thy side,
Nor silk-haired mermaid, with her glass, to guide
Thee through the deep, while, threatening overhead,
Some giant iceberg its dark menace spread,
Like Polyphemus? Or, from lightning shaft,
What fairy shield overhung thy fragile craft?
Or did old Neptune, with his triton train,
When thou wast sinking, buoy thee up again?
Nothing of these? Why then, indeed, '*tis* true—
Heav'n and courage help the feeblest through.
It is a gracious omen here that thou,
Like a new rainbow, with a hope as new,
Art anchored on our shore. It tells of peace.
When jealous anarchies and wars shall cease,
And gallant hearts their direst foes subdue,
United in one cause, Red White and Blue.

Welcome again, " wee craft," to British land ;
Welcome the tiny crew that trimm'd your sails—
Your names with us shall not be writ in sand,
But on our hearts, while memory's tide prevails
Your coming tells this *truth*, scarce understood,
That nothing is impossible that's *good*.

EDWARD FITZBALL.

Notting Hill, September 21, 1866.

———

From " La France," Paris, Oct. 22, 1866.

CHRONIQUE DU SPORT.

Hier, nous avons eu la bonne fortune de nous trouver en compagnie du
capitaine Hudson et de son équipage. L'équipage au grand complet se
compose de M. Frank E. Fitch.
M. Frank E. Fitch est naturellement le *second*.
Le capitaine Hudson est le capitaine du *Red White and Blue* un trois
mâts en miniature construit par M. Ingersoll. Ce petit navire est un chef-
d'œuvre de construction maritime, c'est le bateau de sauvetage par excel-
lence. Avec un pareil bâtiment, il n'est pas de navire en détresse que
deux hommes résolus ne puissent accoster même par la plus grosse mer.
MM. Hudson et Frank E. Fitch l'ont prouvé par une expérience
décisive.
Horace attribuait une triple enveloppe de bronze à l'homme assez intré-
pide pour affronter la colère de l'Océan.
M. Hudson et son équipage n'ont opposé aux lames, souvent furieuses,
qu'une mince carcasse de fer : les bordages du *Red White and Blue* ont
deux millimètres d'épaisseur.—*deux millimétres !*—Pendant trente-sept
jours ces deux hommes ont lutté, travaillé, rêvé, dormi,—dormi, entendez-
vous bien,—dormi de ce sommeil calme et paisible du marin, du marin qui
n'a pas de temps à perdre et qui ferme les yeux avec la conscience d'avoir
bien rempli son devoir
Le voyage de M. Hudson, ne fût-il qu'un simple tour de force, mériterait
d'être classé parmi les plus curieux exemples d'intrépidité, parmi les traits
de courage qui font le plus grand honneur à l'espèce humaine. Mais le
dévouement de M. Hudson a son utilité évidente ; son voyage n'est pas
seulement un acte de courage, c'est un acte de haute philanthropie, c'est
un service éminent rendu à l'humanité.
Que de milliers de créatures humaines arrachées à la mort, grâce aux
instruments de sauvetage dont la collection n'est pas la moindre merveille
de l'Exposition universelle ! Le *Red White and Blue* complète cette admi-
rable collection.
L'expérience décisive de M. Hudson, a été déjà racontée par les jour-
naux d'Angleterre et d'Amérique. Le capitaine Hudson a remis à notre
confrère, M. Lomon, une série de notes et de documents qui accompagnent
son *livre de loch*. M. Lomon publiera bientôt l'histoire détaillée du voyage
entrepris avec tant d'audace et accompli avec tant de succès. Ce sera
certainement une œuvre des plus intéressantes.

———

From the " Weekly Nation," August 23, 1866.

The miniature ship Red, White and Blue has successfully crossed from
New York to London. The distance is not far from thirty-five hundred
miles, and the weather during the voyage was severe ; but the time

required was only thirty-eight days, or *less by one third* than is frequently occupied by vessels of large size. The enterprize was not, it seems, such a mere piece of criminal foolhardiness as might at first thought be supposed. It was performed by men who had the greatest faith in the sea-going qualities of Ingersoll's metallic Life-boat, and were willing to prove them in this way. It is certainly a kind of puffery heroic enough to please Mr. Carlyle, who, on general principles, is not, we suppose, favorable to advertising agencies. It is observed by a daily paper, that the trip of the Red, White and Blue was being made at the same time with that of the immense vessel which was laying the telegraph cable; and the writer remarks that the one voyage "may fairly and contrastively complement" the other, the one being "massivity made nautically available, while the other is dwarfish symmetry rendered demonstratively seaworthy."

From the " New York Evening Post," August 21, 1866.

LIFE-BOATS AT SEA.—The ship Red, White and Blue, which recently crossed the ocean, is an Ingersoll Metallic Life-boat. The result will be regarded as a triumph of American skill and courage which the English, our only competitors, who take great pride in their life-boats, are expected to surpass, or give up the claim for superiority to the Americans. The English may have one that can cross the Atlantic, *but we have yet to record the feat.*

From the " London Standard," November 1, 1866.

The Prince Napoleon, yesterday, honored with an interview, at the Clarendon Hotel, Captain Hudson, of the Red, White and Blue, who explained to his Highness the various incidents connected with the passage of the miniature craft across the Atlantic Ocean. Captain Hudson was accompanied by his agent, Mr. Nimmo.

Paris Correspondence of the " London Spectator."

AMERICAN YACHTING CHALLENGE IN PARIS.—Captain J. M. Hudson, of the Red, White and Blue, who is here, offers a challenge to any yacht in the world, from 250 tons downwards, to sail back to New York, allowing the difference of time for tonnage, both vessels to take the same route, and the stakes to be the same as the Atlantic Race. The difference of time to be allowed to be left to the decision of any yacht club or nautical committee in England.

From the " London Morning Star," September 6, 1866.

Perhaps the most interesting and surprising object ever submitted to the inspection of the public is the wee craft which has accomplished the marvellous feat of safely crossing the Atlantic.

From the " New York Express," August 21, 1866.

If John Bull has sent us the biggest ship ever built, we have sent him the smallest—a perfect beauty—and we don't believe he can reciprocate the compliment. The Red, White and Blue is a pure Metallic Life-boat, and therefore not a show vessel alone, but for use, and is the identical one for which young Ingersoll received the large Gold Medal from the

American Institute in 1865, his motive in building her being the philanthropic one of showing what skill in craft could accomplish.

From the " London Daily Telegraph," August 27, 1866. .

Any thing like pluck and endurance is so congenial to the Anglo-Saxon temperament that no apology need be offered for amplifying the details already known of an undertaking attempted more than once before, but now for the first time successfully achieved.

From the " Dover Chronicle," March 6, 1867.

THE RED, WHITE AND BLUE.

An exhibition—for it is an exhibition—which many thousands have visited the Crystal Palace to see, was on Saturday last brought home to us, to our very doors, as it were; we allude to the little marvel, the wonder and doubt of all sailors, the small American "ship," the Red, White and Blue. Captain Hudson, his "crew" (consisting of his mate, Mr. Fitch), and one other gentleman sailed into Dover on Friday night last, and, after having seen the craft safely moored, took up his quarters at Mr. Ripsher's, at the Providence Inn, in Council House Street. The Providence is, we should say, the only inn in Dover the proprietor of which can say that at a moment's notice he had accommodated the entire "crew" of a three-masted ship, perfect in every thing. The Red, White and Blue is a model of symmetry, a beautiful "thing of life," but, taking her only by her looks, almost too fragile for use. Her tiny masts and spars, ropes, apparently not much stouter than twine, her sails as delicate in appearance as if they were made of cambric, and the nicely rounded and painted hull, with the smooth, clean, miniature deck, cock pit and saloon, lead one to the conclusion that under a glass case would be the most proper place for her. " That gingerbread thing cross the Atlantic," more than one of the nautical frequenters of the pier was heard to exclaim; "psha, I'll never believe it. She might have come across, but it was on deck of some other vessel." We do not intend to enter on this question, as it has been sufficiently argued and proved already; but we are able to state that many of those who held that she was only fit to look at, were completely dumbfoundered and surprised to see her, in spite of all their prognostications to the contrary, sail gallantly out of the habor on Sunday morning, at half-past nine, on her way to Havre, to ascend the river Seine, to be the observed of all observers at the forthcoming Paris Exhibition.

It was not intended, when the Red, White and Blue left Gravesend, to put into Dover at all; but on the representations of Mr. Hamilton, whose knowledge of the harbor enabled him to pilot her in, Captain Hudson gave the Dover people a treat in looking at her. At half-past nine on Sunday morning, to the astonishment of every one, the tug Palmerston took her in tow, and proceeded to escort her beyond the Admiralty Pier; in doing so, the harbor tug shipped a very heavy sea, which filled her decks, compelling her to stop her engines, as there was a perfect gale from the eastward blowing at the time. The sea was running very strong, but the gallant little ship rode the waves like a duck, and on rounding the pier, after being released from the tug, she hoisted her fore topsail and main topsail, steering her course for Havre.

Her length is 26 feet, tonnage 2 tons. Her captain and mate are two quiet looking, determined men, who evidently knew the task they undertook in bringing so small a craft across the Atlantic, and were resolved to carry it out. Much doubt and speculation was created among the Dover pilots and other seafaring men as to the possibility of her having accomplished a task of such magnitude; but this was entirely dissipated on Sunday morning; and those who had been most sceptical freely acknowledged their belief that—from the specimen of what they had seen of her action on the water, she had sailed from New York to England.

From the "South London Chronicle," October 1, 1866.

THE CRUISE OF THE "RED, WHITE AND BLUE."

Afloat, afloat
In a little cock-boat,
And away o'er the wide blue ocean:
Mid waves Atlantic
Careering frantic
With a ceaseless, rolling motion.
Three they were in her; two men who dared
To battle mid-ocean's billow,
And their good dog Fanny went with them and shared
Their narrow and comfortless pillow.

From the land that loomed
On the eyes else doomed
Of Columbus, in days of yore,
The little boat started
Right fearless-hearted,
To visit Old England's shore.
Could those Spaniards have dreamed, as their perils they told
In sailing that shore to discover,
That this terrible ocean would ever behold
Such a cockle-shell boat crossing over?

Lost amid a world of waters,
Foaming breakers all ahead,
Sheets of spray their craft that menaced,
Weltered decks and dripping bed;
Strange and lone the venturous twain
Must have felt on that wide main.

Or when one, in silent watches,
Steered their course with anxious eye,
He the only wakeful being
Underneath the darkened sky;
Solemn must his thoughts have been,
As he viewed that mighty scene.

When the phosphorescent light
Which along their track would play,
Paled before the sunrise bright,
How they must have hailed the day!
One more night of unrest o'er,
One more day the nearer shore.

Nearly forty days they sailed,
.With no comrade save the dog;
But their courage never failed,
  Or in calm or storm or fog.
      And the " Red, White and Blue,"
      Like a brave little craft,
      Through the billows she flew,
      At the breezes she laughed.
      She shook her white sail
      In the teeth of the gale,
      She righted again
      From shock or from strain,
      She parted the spray
      As she darted away,
Like a right little, tight little craft.

      When a whale came near
      They were forced to steer
From his giant presence their bark;
      And once for a day,
      In derision or play,
They were followed along by a shark.
Thrice only, in their venturous trip,
They hailed and spoke a passing ship;
So drear and lonely must it be,
Upon the mid-Atlantic sea.

      And when at last,
      All perils past,
They joyfully steered up the river,
      The companion true,
      Of that scanty crew,
She bade them farewell for ever.
The constant wet she could not stand,
That even their stout endurance tried;
And so in very sight of land,
      Poor Fanny died.

But Captain Hudson and Fitch his mate,
  To the shore they safely drew,
      Amid the acclaim
      Of friends who came
To welcome the daring crew,
And to see the boat that bore them o'er,
  The Ingersoll Life-boat, tight and true.
Now here's three cheers for the tiny craft—
  Hurrah for the " Red, White and Blue."

ADDITIONAL EVIDENCE.

From the "New York Herald," April 29, 1866.

STRUGGLE FOR LIFE—THRILLING SCENE.

Burning of the Steamer City of Norwich—The Boats of the Steamer Useless—Eleven Lives Lost—The Balance of the Passengers Saved by the Propeller Electra with Ingersoll's Metallic Life-Boat.

The affrighted men and women in the water, some with life-preservers, and others floating on boxes and boards, now made for the Life-boat belonging to the steamer; and in an instant she was sunk to the level of the water by the number of those who endeavored to crawl over her sides. The captain and the pilot did all in their power to keep the crew in subjection, and, as they struggled and fought to get into the boat, begged them to hang by the sides, and not get into her. * * * * But words were unavailing. The men tore at each other like madmen, and struggled to raise themselves into the already filled boat. At last, struck by a heavy wave, the boat swamped, and everybody was thrown into the water. *All efforts to right her proved fruitless*, and during those awful *few moments more than one went down to rise no more.* The scene that was now presented by the burning vessel and her surroundings was terrible in the extreme. The flames from the City of Norwich lit up the Sound for miles around, revealing a most appalling sight. For over a quarter of a mile on every side the waters were covered with burning boxes and bales of all descriptions, amid which the struggling, drowning ones, the swimmers, and those with life-preservers, were endeavoring to sustain themselves.

THE RESCUE.

At the moment the flames first shot up through the decks of the Norwich, the propeller Electra, from Providence to New York, was about a mile astern of the doomed vessel. As the whistle sounds of distress reached her, and the flames broke out on the Norwich, Captain Nye turned the bow of his vessel toward the scene of disaster, while the men were put to work in cutting loose the two Ingersoll Metallic Life-boats on the decks. Being up within an eighth of a mile of the burning vessel, the Electra was stopped, as it was feared the floating mass of burning boxes and cases would set the latter on fire, should she continue to advance. The two boats were manned and lowered, and sent off to the rescue of the men and women struggling in the water. A third Ingersoll Metallic Life-boat was also got in readiness and shoved off by volunteers from among the passengers. These three boats rescued every person that could be found in the water, and returned to the Electra in safety. As each life-boat arrived with its human freight, snatched as it were from the jaws of death, Captain Nye, to whom too much praise cannot be accorded, assisted by Mr. Edward Bynner, the Agent of the Boston and New York Steamboat Line, and the passengers of the Electra, did all that was possible under the circumstances to make things comfortable for the rescued passengers and crew of the Norwich. Mr. Tracy, the pilot, said he had been a seaman for thirty years, and had seen many sorrowing sights, but those attendant upon this disaster eclipsed them all in its appalling horrors.

LETTER FROM SURGEON JAMES SUDDARDS, ONE OF THE SURVIVORS OF THE ONEIDA HORROR.

Thrilling Story—Wooden Boat Broken in Pieces—Wooden Boats will Shrink and become Rotten—No Dependence in the Dark Hour of Shipwreck and Peril—Solemn Words.

We will not devote more space in giving instances of the success of the Ingersoll Life-boat. They never have failed to perform their life-saving mission. We here most solemnly wish to say to ship owners and navy officers, that wooden boats cannot be relied upon in the dark hour of wreck and calamity on shipboard. Wooden boats will shrink and become rotten, and a blow or an explosion of a gun that would shatter a wooden boat for all practical purposes will not affect a metallic boat. As a matter of economy they are also desirable. Once bought, that is an end of the expense until the ship itself wears out. We will here give an extract from a letter from Surgeon James Suddards, of the steamer Oneida, to his father, Rev. William Suddards, D. D., dated " Yokohama, January 31, 1870." It is like one speaking after having been suspended over the abyss of a watery grave. He says: " Out of a personnel of 25 officers and 150 men, 9 officers and 54 men are left to tell the tale. We were still hanging at the davits when the ship began to roll in that peculiar way which precedes foundering, and the boat was dashed against the side of the ship, threatening to dash her in pieces. I looked on the deck and saw no one about, and gave order to lower away and hang by the falls. The fall got jammed, and had to be cut away with a knife. Had we been three minutes longer at the davits it would have been too late, as she went down like a shot after starting, and the suction would have carried our boat down with the wreck. I may mention here, that when the boat was brought up to the Idaho, *she nearly sank alongside ;* and on examination it was found that *seven pieces were broken on her starboard side, and one of the planks was knocked an inch out of place.* This must have been done by *striking the side of the ship,* and convinces me that we could not have saved any more in our boat, as she *would have filled and gone down with a heavier load."* It is needless to add, had this boat been metallic, it would not have been more injured than would the steam boiler if hanging over the sides, in fact, would not have been injured in the least, and many more valuable lives been saved. All naval vessels should certainly have a reasonable number of good metallic Life-boats. We are now engaged in supplying an important European government with our Life-boats for their naval vessels.

When the Ingersoll Metallic Life-boat was first introduced to the public, in 1860, the following were a few of the opinions of the Press:

There can be no doubt that many so-called metallic Life-boats are in many circumstances no life-boats at all, but rather the contrary. Mr. O. R. Ingersoll, whose Boat and Oar Bazaar in South Street is really one of the peculiar institutions of New York, has invented a new Boat on an entirely new principle. We should like to see it on every passenger vessel plying to and from this city—*New York Express.*

It has the additional merit of being much cheaper than other Life-boats.—*New York Commercial Advertiser.*

The business streets were startled Wednesday, by the exhibition of a New Metallic Life-boat, the invention of Mr. O. R. Ingersoll, who has made his name a household word among Aquatic Amateurs by his Race and Sail Boats.—*New York Times.*

INGERSOLL'S

METALLIC LIFE RAFT.

We have just completed the construction of a new

METALLIC LIFE-RAFT.

We claim for it:

1st. It is fire, worm, and rat proof.

2nd. Its air compartments are so arranged, that were even one half of them stove, the Raft would still sustain *all the people that could crowd upon it.*

3rd. It can be thrown off the side of a vessel, and it is *immaterial which side comes up.*

4th. It is *impossible to sink or swamp it.*

5th. From its peculiar construction, it will last as long as any ship—*always air tight* and in good order. A large number of them, sufficient to save all the lives that any vessel can carry, can be stowed in a very small space.

6th. It is offered at a *less price* than any other raft, notwithstanding its great *superiority* and *durability.*

New York, April 7, 1870.

From the "New York Commercial Advertiser."

BOATS—THEIR SIZE, USE, VARIETIES, EARLY HISTORY, &c.

Whether man first took the idea of boats from an insect that floated across a rivulet on a leaf, or a beaver carried down a river upon a log, or a bear borne away upon an iceberg, we cannot say; but it is highly probable that he first learned from animals, whose natural element it is, the manner of supporting himself upon, and forcing his way through it. To this day, the Indians of our own country often use the primitive means of boating, and cross a rapid stream by clasping the trunk of a tree with the left leg and arm and propelling themselves with the right. Thus the first step was taken; the second was either to place several logs together, thus forming a raft, and raising its sides, or to make use of a tree hollowed out by nature. The next step was to hollow out by art a sound log, thus imitating nature.

The early navigator, seated in his hollow tree, might first seek to propel himself with his hands, and artificially lengthen them by a piece of wood fashioned in imitation of a hand and arm—a long pole terminating in a thin flat blade. Here was the origin of the modern row-boat, one of the most graceful inventions of man. For centuries the art of boat building did not advance as rapidly as some of the kindred arts, and men were wont to trust themselves in rude, clumsy and unsafe vessels; but during the last century, the art of boat building has progressed with remarkable rapidity, and New York has taken the lead in this branch of business. At the present time there are in this city ten firms engaged in boat building, and the yearly consumption of wood used in the construction of boats is about 300,000 feet of white cedar, and 150,000 feet of white oak. These materials must be of the very finest quality. Each of these shops employs an average of three men, with the exception of Ingersoll's Bazaar, where 100 to 300 men are employed. This is the largest boat building establishment in the world. In nearly every port in the world are to be found New York built boats. They are of almost an endless variety of model, and many are fashioned after some fancy of their owners: but there are a large variety that have a standing model as well as name.

First comes the yawl, which is from 10 to 22 feet in length, with a beam in proportion. They are used mostly by our coasting craft. Stern boats have their peculiarities, and are used by all classes of vessels; these are from 16 to 20 feet in length. The workmanship is much finer than that of the yawl. Long boats, or lighters, are built for ships, and are carried upon their decks, to be used for transporting cargo in ports where vessels cannot lie at the docks. They are very strongly built. A Moses boat is one in size and model between a yawl and a long boat. Wash-board boats are next in the list, and are propelled by from four to six oars, are both double and single banked, and vary in size from 16 to 24 feet. They are much used by our men-of-war, being considered indispensable in a rough cross sea; the wash-board preventing much water from coming over the gunwale. Double-headers, or surf-boats, are sharp at both ends, and are very valuable in landing in a surf or saving life through a heavy sea. They are much used on the coast of Africa, and other places where another form of boat would imperil the occupants by a liability to capsize while passing through the breakers. They are from 16 to 30 feet in length, being very broad. Another class of boats, called the African surf-boat, is now being extensively built. These are very large and heavy, of about four tons burden, and are used in taking in cargoes of palm oil, and often negroes. They are about 30 feet in length. Pilots' yawls are constructed expressly for that service, and although light are extremely strong. They are uniformly 16 feet in length. Hell gate pilot's boats are also adapted to that peculiar service, and are of beautiful model. They are clinker built, copper fastened, and are usually 14 feet in length. Captains' gigs are four, six and eight oared, and are of a fine lean model, and more expressly designed for speed than carrying capacity. They are very light, and vary in length from 18 to 26 feet. The fittings of some of these boats cost from $20 to $60, and consist of brass row locks, rudder yoke, gratings, awnings, &c. Whitehall boats, used for the transportation of passengers from shore to vessels lying in the river, &c., are usually about 19 feet in length, and will comfortably carry six persons. These boats are much used in our waters.

Ladies' wherrys are a style of boats lately introduced by Mr. O. R. Ingersoll, and are intended for ladies' use exclusively. Quite a number have been built this season for some of the ladies of our first families. The boats are very light, weighing only about 60 pounds, clinker built and copper fastened, very graceful, and are from 10 to 12 feet in length. Pleasure boats, much used by gentlemen who have residences near the water, are usually from 10 to 20 feet in length, and copper fastened. Sail boats are built after various models, and are both smooth and clinker built. Gunning boats are mostly built for the Southern trade, and are extensively used in the swamps at the South. They are built with a portion of the deck arranged so as to screen the sportsman as he drifts towards his game. They are very light in weight, and drawn very little water. They are about 18 feet in length. Scows are of the primitive style of boats, and, at the present time, much used upon our small lakes and rivers. They are the boats which usually accompany rafts, &c., and are about 12 feet in length. Fishing boats are constructed of great width, are very buoyant, and are used for fishing on banks and shoals. They are intended for one man, and rank among the best of sea boats, and with careful management live in the worst description of sea. They are clinker built, and are from 10 to 12 feet in length. Skiffs are an improvement on the primitive scow, being sharp at one end. They vary in length from 12 to 16 feet. The batteau is another of the scow species, and is used in the neighborhood of rapids. They are about 12 feet in length.

RACE BOATS.

At the present day a very large business is done in the building of race boats, and much enterprize is exhibited by the several builders in constructing fast craft. The race boats of New York are conceded to be far superior to those built elsewhere; hence most of the orders for that class of boats are sent to this city. First in this class is the lap streak race boats, which are in use throughout the country. They are from 30 to 55 feet in length.

The club boat is used by amateur clubs, and embraces both lap streaks and smooth build, varying in length from 26 to 36 feet, and propelled by from four to six oars. Among the latest inventions in this class of boats is the mahogany shell boat. These are the lightest that float, excepting the birchen canoe of the Indian; and so frail is their construction and liable to capsize that the utmost precaution is necessary on the part of the occupant to prevent such an occurrence. They are about 30 to 55 feet in length, from 18 to 22 inches in width, and so narrow that the sculls are used in outriggers, projecting from either side about two feet. They weigh from 75 to 80 pounds.

———

STEAM YACHTS.

From the " New York Express."

TRIAL TRIP OF THE STEAM YACHT "MINNIE."—This yacht made her trial trip on Tuesday—starting from Ingersoll's boat establishment—making a trip around the harbor through the Sound. The yacht is 42 feet in length, and is of beautiful finish and style. She made fifteen miles per hour on the trip. No expense has been spared to make her cozy and unique. As she passed up the river she was hailed by all the steam whistles on steamers and factories.

_____

From " Frank Leslie's Illustrated," March 5, 1870.

IRON CUTTERS FOR UNITED STATES SERVICE.—Quite a novelty in the line of steam navigation are the powerful little iron vessels recently constructed in New York city for the revenue service about Alaska. Their dimensions are thirty-seven feet in length, ten feet four inches beam, and five feet in depth. They are furnished with engines of eighteen horse power, which are capable of making two hundred and fifty revolutions per minute. On a trial trip they made an average speed of eleven knots per hour, which exceeded the expectations of the constructors. The vessels were built at the Ingersoll Works, 159 South Street, New York, and more than answer the demands of the revenue service.

_____

From the " New York Leader."

NEW SAIL AND STEAM YACHTS, ROW AND SHELL BOATS.—During one or two visits to Mr. Ingersoll's well known establishment in South Street, we saw a large number of yachts and row boats, among others the following : The jib and mainsail sloop " Empress," which was shipped to Mobile some time since. Since her arrival out there she has won every purse for which she sailed, and is now champion of Mobile bay. The cat-rigged boat "Water Lily," recently sent to St. Paul, Minn., is a handsome boat, twenty feet long and eight feet beam. Mr. Ingersoll has now on the stocks a steam yacht of eighteen tons; she is forty-five feet long, eight feet six inches beam, and five feet deep. The steamship " Quaker City," which is now making an extended ocean excursion, carried a shell thirty-six feet long, sixteen inches and a half wide, also a 14 feet sprite sail boat, both of which are to be used in the various harbors visited by her. The four-oared club boat " Minnehaha," recently completed for a boat club, was well worthy of inspection. She is made of white cedar, with mahogany trimmings.

_____

From the " New York World."

INGERSOLL'S YARD—METALLIC LIFE-BOATS.—Mr. Ingersoll is building a large and peculiarly modeled government vessel for one of the Central American Republics; and having lately introduced a novel and valuable Metallic Life-boat, he is engrossed to his utmost abilities in building them for nearly all the steamers in course of construction.

_____

REGATTAS.

EMPIRE CITY REGATTA.—The six-oared boat "Gulick," that won the champion race at the above regatta, was built by O. R. Ingersoll.

_____

From the " New York Sun."

BOAT RACE.—A spirited boat race came off on Friday between the " Lucy," rowed by the brothers Donahue, and the " Elizabeth," rowed by Roder and Paterson. The Lucy won the race in forty minutes, said to be the quickest time on record. She was built by Ingersoll. The Elizabeth is better known as the Burns Champion Boat, and was considered the fastest boat in New York. The betting on her was two to one previous to the race.

REGATTA AT YALE COLLEGE.—At the recent regatta, the six-oared boat Varuna, built by Ingersoll, took the first prize. Four boats contested.

BOAT RACE IN CLEVELAND.—The match race between the boats "Mazeppa" and "Rover," at Cleveland, Ohio, on the 25th of September, resulted in favor of the Mazeppa, built by Ingersoll, of New York.

From the "New York Clipper."

Clear the track! Messrs. Ingersoll, the world wide celebrated boat-builders, have this week sent from their bazaar a four-oared race boat, "Addie," six-oared race boat, "Independent," for Dobbs Ferry, and have laid the keel for the six-oared race boat for the noble lads of Yale.

From the "New York World."

RACE, CLUB AND SAIL BOATS—THREE RACES WON IN ONE DAY.—The amateur rowing clubs are preparing for the season with unusual activity. The Ydrad boating association of Cleveland, Ohio, have ordered from O. R. Ingersoll a six-oared club boat, which will soon be completed. Her model is one of the finest ever built in this city. She is forty-five feet long, three feet beam, and will weigh about two hundred pounds. She will have outriggers, and will carry a coxswain. She will be furnished in the most costly manner, without regard to expense, it being the ambition of the club owning her to have a boat superior to any boat of her kind on the western waters. The boat reflects much credit on both her builders and owners. The Pioneer boat club of Albany, N. Y., are also having a splendid four-oared barge built by Mr. Ingersoll. This club is one of the crack clubs of the State. Mr. Ingersoll is about to build other boats for different clubs in various portions of the country. His boats are widely known, he having furnished all styles of rowing craft to the clubs at Albany, Pittsburgh, Rochester, Cleveland, Harvard College, Savannah, Mobile, New Orleans. At the regatta of the American Institute off Castle Garden, the boats built by him *won the first prize in three races.* The Troubler won the sailboat race, the George W. Chapman the four-oared race, and a nineteen foot boat won the prize in third class. The celebrated nineteen foot working boat, Henry Stork, which won thirty races in eighteen months, and the champion six-oared boat, Gulick, of the Aurora club, came from his establishment.

From the "Williamsburgh Times."

GOVERNMENT LAUNCHES—QUICK WORK.—O. R. Ingersoll, the South Street boat builder, has just completed a large number of launches for the government. They were built in sixteen days, are of three sizes—40 feet long, 12 feet beam; 37 feet long, 10½ feet beam, and 34 feet long, 9 feet beam. They are now being put on shipboard. The ship Planter has removed some of her deck beams, and is to carry forty of them. Such an enormous amount of work done in sixteen days is an unparalleled feat, and seems almost an impossibility. Three hundred men and several steam saw and planing mills were brought into requisition.

From "Porter's Spirit," New York, June 4, 1861.

We attended, last week, the annual exhibition of Mr. Ingersoll's Boat Bazaar, in South Street, and were much pleased with what we saw. These annual displays have been kept up for many years by father and son, the latter having now succeeded his senior, Mr. C. L. Ingersoll, who was so many years the patron and promoter of regatta and boating amusements in our city.   In fact, those who can retrace our rowing contests and yachting sport to some twenty years past, must, as we cheerfully do, identify Mr. C. L. Ingersoll's name prominently with all our early aquatic events.   The present mammoth establishment of O. R. Ingersoll is well worthy a visit at any time from those who take an interest in viewing well modeled boats and yachts.   The various floors are devoted to the making of boats of every variety, consisting of Launches, Surf Boats, Captains' Gigs, Club Boats, Racing Shells, Outriggers, Yawls, Dinkeys, &c., in short, every description of boat known to American waters, including Yachts of all sizes, from the Pilot Boat to the little Model Yacht.   We noticed among the boats on exhibition two beautiful club boats, several single and double scull boats, and a sloop and yacht just completed, were deserving of notice, as creditable specimens of Mr. Ingersoll's skill.

From the " New York Evening Express."

Lovers of aquatic sports, and of all that is beautiful and fairy-like in the boat line, have still a chance to enjoy that rare treat, which is only afforded once a year, of paying a visit to the great boating fair at Ingersoll's Bazaar in South Street.   Here are to be seen all varieties of boats—yawls, life boats, yachts, sail boats, club boats, dingys and cutters; gigs and quarter boats; skiffs and whale boats—no kind of boat a man may want to see, or, still better, may want to buy, but can be found at this beautiful fair.   It was thronged last week; and we presume that many thousands will visit it before Saturday night next.

From the " New York Clipper."

We dropped into Ingersoll's Boat Fair a few days since.   We were much surprised, and could hardly believe it possible that so many varieties of boats could be gathered together in one establishment.   Yet we beheld yachts, sail boats, club barges, row boats of various patterns, skeleton boats, working boats, surf, whale and life boats ; ships' yawls, ducking boats, and in fact many others, the names of which escaped our memory, and these, too, all under one roof.

From the " New York Herald."

NEW SURF BOAT.—Mr. O. R. Ingersoll, the boat builder, has shown us the model of a surf boat that he has invented for the purpose of landing any body of men through the heaviest surf, without inflicting any damage to it.   Its peculiar construction does not render it liable to any of the accidents that ordinary surf boats are subject to.   It has a flat keel, or, we might say, no keel, and is so constructed that when stranded it strikes in the middle.   It is strongly built, and braced with wrought iron bands in different directions.   It has been thoroughly tested, and is a great success.   Such a boat would be very useful in naval expeditions, in landing troops.

From the " New York Evening Post."

Our reporter yesterday visited Ingersoll's Boat and Oar Bazaar, the largest establishment in the United States, and, probably, in the world.   Upon its floors are congregated some of the daintiest specimens of the smaller sea-going craft that the hand of the boat builder ever fashioned.   The first thing that strikes the visitor's eye upon entering are a couple of race boats, with keen sharp look about their prows, that shows they are evidently destined to make their way through the waters.   Farther on are gayly painted yachts, combining all the requisites of strength, beauty and speed. The barges, on the same floor, are justly celebrated for their build and finish. These boats are sent to every quarter of the globe, and doubtless, at the present day, furrow the waters of every sea. The establishment is a real curiosity shop, filled with all manner of little sharp nosed, shovel nosed, and snub nosed craft of his own make, and a general assortment of little nautical wonders from various parts of the world. In the basement are piles of ashen oars, lithe and slender, for the little racer, and broad and strong for the becalmed sloops and schooners on the rivers. In the upper stories there is a perpetual din and clatter of tools, hammers, mallets and saws. Around, rows of skeletons, like umbrella frames, and the bones of horses and elephants.

From the "Boston Traveller."

Ingersoll has won a reputation as wide as the world. His boats are to be met with from Hudson's Bay to the China Seas, and every where have given universal satisfaction.

From the "New York Journal of Commerce."

A novel exhibition is afforded by Ingersoll, who displays something like two hundred boats, of every description, decked out in flags and other adornments. The exhibition rooms comprise five floors of the building.